INDUSTRIAL WATER QUALITY REQUIREMENTS

N.MANIVASAKAM M.Sc.,
Formerly **CHIEF WATER ANALYST**
PRINCIPAL PUBLIC HEALTH LABORATORY
COIMBATORE-641018 (T.N.)
INDIA

Chemical Publishing Company

INDUSTRIAL WATER – QUALITY REQUIREMENTS

ISBN: 978-0-8206-0004-8

Chemical Publishing Company:
www.chemical-publishing.com
www.chemicalpublishing.net

Originally published 2005

Printed in the United States of America

DEDICATED

TO

THE LOTUS FEET OF

LORD MURUGA

I gratefully acknowledge,

1) The Resources Agency of California, State Water Resources Control Board, U.S.A

2) World Health Organization, Geneva, Switzerland.

3) American Society of Mechanical Engineers, New York

4) American Society for Testing and Materials, Philadelphia

for their kind permission to reproduce some of the tables, and paragraphs from their publications,

- N.MANIVASAKAM

PREFACE

It is well known that without water no life can exist on earth. Similarly without water, many of the industries cannot exist as water is used in multifarious ways in industries. Water is the raw material, water is the ingredient, water is the medium of reaction and water is the conveying medium. Besides, water is used to generate steam and it is the major cooling agent. Without water, washing operations cannot be carried out. Needless to say that each process requires a specific quality. While every industry is keen on the availability and abundance of water, it is a sad fact that much importance is not bestowed upon the quality of water. Adherence to the quality requirements of water would result in a better product, protection of equipments, increased efficiency and avoidance of losses. Even if the industry is much interested to know the quality, there is a dearth of literature and even the available literature is scattered. The intention of writing this book is to provide comprehensive information on the quality requirements of water for each industrial use in a single volume. Almost all the major wet processing industries and quality requirements of water for such industrial processes are described in this book in detail.

For easy understanding this book is divided into five parts, namely, 1) Introduction, 2) Impurities present in Water and their Effects on Industrial Uses, 3) Industrial Waters and their Quality Requirements, 4) Drinking and Irrigation water and 5) Impurities present in Water and their Effects on Water Treatment Systems,. Every effort is made to provide all the possible information through these sections. Chapters are arranged in alphabetical order for easy access to the required chapter.

This book would fill the long felt need of the availability of such an one volume scientific discussion and hoped to be a boon to all the industries, work managers, plant managers, water treatment personnel, chemists, and to water treatment consultants, firms engaged in water treatment programs etc. This book would also serve as a ready reference to students of Industrial Engineering, Chemical Engineering, Industrial Chemistry, Applied chemistry, and to various other faculties of Colleges and Universities.

The preparation of this book is possible only with the help and assistance of a number of people. Their assistance made this book a most useful volume. The author owes a great debt of gratitude to all of them. Some of the organizations and publishing houses have been most gracious in granting permission to reproduce certain tables and paragraphs from their publications. Their cooperation is gratefully acknowledged. A special debt of gratitude is due to M/s. Chemical Publishing Company, Gloucester, USA, for their keen interest shown towards the publication of this book.

Comments and criticism as well as suggestions for the improvement of the book are solicited.

Coimbatore – 42 (TN)
India.

- N.MANIVASAKAM

This book is divided into the following parts.

CONTENTS

PART III INDUSTRIAL WATERS AND THEIR QUALITY REQUIREMENTS

PART I

INTRODUCTION

CHAPTER - 1
INTRODUCTION

Water is the elixir of the life and without water no life can exist on earth. Water is not only the fluid that sustains life but it is invariably used for a variety of purposes. Needless to say that water is one of the important materials used in industry. The major industrial uses of water are steam generation, cooling, washing, conveying of materials etc., Above all, it is the major raw material, which becomes an ingredient in the finished product. This apart, it is used in a number of ways in industry.

The industrial progress is entirely dependent on water. Besides the availability of adequate quantity, water should also conform to certain quality requirements to suit the industrial needs. Pure water does not contain any impurities. But the availability of such waters are very rare. The contaminants change the water quality, and the quality requirement of water for any use is interpreted in terms of these impurities. This book delves with such impurities and their effects on industrial uses.

Water is relatively pure when formed, but when it enters to atmosphere it absorbs gases, particularly oxygen and carbon dioxide. Upon reaching the earth, it dissolves the mineral matter with which it comes in contact. The amount of dissolution depends on various factors such as the solubility of soil or rocks in the area and the length of time of contact.

The common minerals encountered in natural waters are carbonates, bicarbonates, sulfates, chlorides and nitrates of calcium, magnesium, sodium and potassium. In addition to these minerals, silica and metals such as iron and manganese and also present. Small amounts of phosphates and fluorides also occur in all waters. Practically all waters contain gases (mainly oxygen and carbon dioxide), absorbed from the air. A little organic matter is also usually present. This apart, most of the surface waters contain algae, diatoms and a wide variety of bacteria and other plant and animal forms. In addition, they also contain solids ranging from sand to colloidally dispersed particles.

Surface waters are generally softer and less concentrated than ground waters. They contain high amounts of dissolved oxygen and sediment. They often show variations in their characteristics, whereas ground waters are characterized by essentially constant composition.

Ground waters are normally high in total dissolved solids, hardness, carbon dioxide but low in suspended solids. Because of the higher mineral matter content, they are less preferable for industrial and domestic uses.

The foregoing is a discussion on the sources and general nature of water. A detailed analysis of water will clearly reveal its characters viz, chemical constituents, physical characteristics, biological contaminants and radiation levels. Acceptability of water for use in industry is based upon these characters.

The term 'Water Quality' is only relative to the proposed use of water. The quality requirements for industrial uses vary greatly with the type of industry and the purpose for which it is used. There is no 'universal water' that can be used for all industrial purposes. Each industry requires water of specific quality that differs industry to industry. A water that is of satisfactory quality for an industrial purpose may be of very poor quality or even dangerous for another industrial use. For example, brewery and bakery essentially require some amount of calcium sulfate in water, whereas water intended for boiler purpose should not contain even a trace of calcium sulfate. There is an axiom, "One man's bread is another man's poison". This holds good for water quality also. Hence it is imperative to have a knowledge on the impurities present in water, their effects on industrial use and quality requirements of water for industrial purpose. In Part II of this book, the impurities present in water and their effects on industrial uses are discussed in detail. Impurities present in water also affect the water treatment systems such as Ion exchangers and Reverse Osmosis units. They are discussed in Part - III. Part - IV describes the quality requirements of water for each industry.

PART II

IMPURITIES PRESENT IN WATER AND THEIR EFFECTS ON INDUSTRIAL USES

CHAPTER – 2
ACIDITY

Acidity is not a specific constituent, and it is a measure of the effects of combination of substances and conditions in water. It may be defined as the power of a water to neutralize hydroxyl ions and is expressed in terms of calcium carbonate.

Acidity is usually caused by the presence of free carbon dioxide, mineral acids such as sulfuric and weakly dissociated acids. Iron and aluminum salts hydrolyze in water to release mineral acidity. Surface waters and ground waters attain acidity from humic acids and from industrial wastes such as pickling liquors, and from acid mine drainage. Most of the natural waters do not contain any mineral acidity and contain only carbon dioxide acidity.

There is no specific limit for acidity but it is indirectly controlled by the limits on pH value. Excessive acidity is detrimental to metallic and concrete structures and should be neutralized to mitigate corrosion. Further, most of the industrial processes need water free from acidity, even from carbon dioxide acidity.

CHAPTER – 3

ALKALINITY

(See also, Chapter - 5 "BICARBONATES AND CARBONATES")

Alkalinity is not a specific constituent but is the quantitative capacity of the water to react with hydrogen ions. The alkalinity of natural or treated waters is normally due to the presence of bicarbonate, carbonate and hydroxide compounds of calcium, magnesium, sodium and potassium. Borates and phosphates and silicates also contribute to alkalinity. Some other ions not ordinarily found in natural waters such as arsenate, aluminates and certain organic anions in colored waters could also increase the alkalinity. Because of the relative abundance of carbonate minerals and because of the ready availability of carbon dioxide that enters into equilibrium with them in water solution, most waters contain bicarbonate and carbonates only. The occurrence of hydroxide ions in natural waters is very rare, unless artificial contamination has occurred. In surface waters where algae are fluorishing the waters contain carbonate and hydroxide alkalinities. Boiler waters usually contain carbonate and hydroxide alkalinities. Chemically treated waters, particularly softened waters (by lime or lime soda process), contain carbonate and hydroxide.

The determination of alkalinity provides an idea of the salts present in water. If the alkalinity is equal to hardness, calcium and magnesium salts only are present. If the alkalinity is greater than hardness, it indicates the presence of basic salts - sodium and potassium in addition to those of calcium and magnesium. If the alkalinity is lower than hardness, neutral salts of calcium magnesium must be present that are not carbonates; usually these are sulfates.

The alkalinity of a water has little sanitary significance. However it is significant in many uses and in the treatment of water and waste water. Alkalinity is an important parameter involved in corrosion control. It must be known in order to calculate the Langelier Saturation Index. Alkalinity and pH are the factors in determining the amenability of waste waters to biological treatment. The alkalinity in excess of calcium and magnesium concentrations is significant in determining the suitability of a water for irrigation.

CHAPTER - 4

ALUMINIUM

Aluminum, the most abundant metal in the earth's crust is present in natural waters through contact with rocks, soil and clay containing aluminum. Aluminum sulfate is the most widely employed chemical for clarification of water. Such clarification process also contributes aluminum to treated water. Wash waters from water treatment process is also a source of aluminum to water. Corrosion and erosion of aluminum tubing in condensers and feed water heaters can also be a source of aluminum in industrial waters.

Aluminum is undesirable in industrial waters because of its sludge forming characteristics. Waters containing aluminium are undesirable for textile processing as they form certain color lakes especially with special dyes such as alizarin. For textile processing it is desirable to keep the aluminium content within 0.1 mg/l to protect from any undesirable effect. The presence of aluminum in water intended for rayon manufacture and dyeing of rayon is also highly undesirable as it tends to form hydroxide soaps that may produces stains. The recommended maximum permissible concentration of aluminum for rayon manufacture is 0.25 mg/l. The residual alum content should not exceed 0.4 mg/l (as alum) or 0.04 mg/l (as aluminum) for rayon dyeing.

Aluminium is reported to cause certain problems for dialysis patients and it may have some effects on general human health.

CHAPTER - 5

BICARBONATES AND CARBONATES

(See also, Chapter - 3 "ALKALINITY")

Bicarbonates and carbonates are the chief sources of alkalinity*. The alkalinity of a natural water normally consists of calcium and magnesium bicarbonates. Some times sodium bicarbonate is also present to a significant extent. (Surface water in impounded reservoirs some time contain hydroxides. During daytime, photosynthetic activities of algae and other aquatic plants utilizes all the carbon dioxide converting the bicarbonates and carbonates to hydroxides).

Like pH, alkalinity is an important parameter in deciding whether a water supply has the tendency to form scales or to corrode. Hence the concentration of bicarbonate and carbonates are very significant.

In boiler water applications, knowledge of the bicarbonate and carbonate of the water is essentials to decide a proper treatment as they may cause some adverse effects. For example, if bicarbonates and carbonates introduced with the feed water decompose as shown below forming hydroxide and releasing carbon dioxide.

$$2HCO_3^- \xrightarrow{heat} H_2O + CO_2 + CO_3^{2-}$$
$$CO_3^{2-} + H_2O \xrightarrow{heat} 2OH^- + CO_2$$

The hydroxide produced in such a way would react with magnesium to form magnesium hydroxide. Besides, the released carbon dioxide will be carried along with steam and dissolve in condensate to form carbonic acid which is corrosive to steam condensing equipment and to condensate return system.

However the presence of carbonate to a certain extent is desirable with regard to waters intended for low pressure boilers, as the carbonates prevent deposition of calcium -

* Other ions such as borate silicate and phosphate also contribute to alkalinity. But in natural waters they are present only in insignificant concentrations. However industrial waters do contain phosphate and hydroxide contributing much to alkalinity.

sulfate which forms a very hard scale. Carbonates in optimum concentration give proper protection to equipment from corrosion by forming a thin layer over the entire surface of the metal.

Presence of bicarbonates and carbonates in water is also undesirable for certain industrial processes such as sugar manufacture, textile and pottery. The recommended maximum limits for bicarbonates for these processes are 100 mg/l, 200 mg/l and 200 mg/l respectively. High bicarbonate contents are reported to cause swelling of skin in tanneries. Bicarbonate if present as sodium bicarbonate is reported to cause a white shell in ice. Excessive carbonates are reported to be detrimental in the production of carbonated beverages also. Carbonates are also undesirable for brewing, the maximum limit being 60 mg/l.

CHAPTER - 6

CALCIUM

[See also, Chapter - 15 "HARDNESS" (TOTAL)]

Calcium salts are among the most commonly encountered substances in water. The presence of calcium in water is mainly due to its passage through or over deposits of limestone, dolomite, gypsum and other gypsiferous materials. Calcium and magnesium salts are the major scale–forming substances in most raw waters.

Calcium is an essential element and human body requires approximately 0.7 to 2.0 g of calcium per day as a food element, the amount which cannot be supplied even by hard waters. Infact, calcium deficiency is the most common nutritional lack in many parts of the world. However waters with high calcium content are undesirable for household uses such as washing, bathing and laundering because of the consumption of more soap and other cleaning agents. Moreover, calcium salts tend to cause incrustations on cooking utensils and water heaters.

Calcium in industrial water is undesirable mainly due to the formation of scales in boilers, heating system and cooling system.

Calcium is the principal scale forming cation. Among the compounds of calcium, calcium carbonate and calcium sulfate are the chief scale formers. Calcium hydroxide too form scales. With silica it forms very hard and adherent scale of calcium silicate. Calcium carbonate scales are softer in nature and easier to remove, while calcium sulfate scale is very hard and difficult to remove. Calcium hydroxide is an abnormal scale-former. It does not occur in natural water and would be introduced into boiler only through improper treatment.

Calcium is objectionable in process waters also. In textile interblent processing, it reacts with alkaline solutions to form precipitates and curds, especially during washing operations using soap. It also causes difficulties in rinsing operations in electroplating. It is reported to affect certain fermentation processes too. The following concentration of calcium have been recommended as limits for various industrial uses.

TABLE – 6.1
RECOMMENDED THRESHOLD CONCENTRATION OF CALCIUM IN INDUSTRIAL WATERS*

Process	Threshold concentration in mg/l
Brewing	
Light beer	100 - 200
Dark beer	200 - 500
Soda and sulfate pulping	20
Sugar manufacture	20
Textile manufacture	10

However presence of calcium sulfate (and calcium chloride) is beneficial in waters intended for brewing. It helps to maintain the acidity of the wort and therefore causes more complete coagulation of albuminous matter. It also reduces the solubility of the bitter substances of the hop.

* Reproduced with the kind permission of the Resources Agency of California, State Water Resources Control Board, California from their publication "WATER QUALITY CRITERIA", 1963, edited by J.E. Mckee and H.W. Wolf.

CHAPTER - 7

CARBON DIOXIDE

[See also, Chapter - 29 "OXYGEN (DISSOLVED)"]

Carbon dioxide is a normal component of all natural waters. It dissolves in water in varying amounts and the dissolution depends on temperature, pressure and mineral content of the water. Polluted waters acquire carbon dioxide by the biological oxidation of organic matter. Surface waters normally contain less than 10 mg/l free carbon dioxide while some ground waters contain 30 - 50 mg/l.

The presence of free carbon dioxide is an important consideration because of its corrosive properties. As far as corrosion is concerned, the most troublesome gas (in boiler operation) next to oxygen is carbon dioxide. In addition to the existing carbon dioxide, bicarbonates (and carbonates) present in most boiler feed waters also yield free carbon dioxide on decomposition, which combines with water to form a weak acid known as carbonic acid. This acid reacts with the boiler metal to form their bicarbonates eating away some metal for such reaction. The formed bicarbonate once again decomposes and liberate carbon dioxide which combines with water to form carbonic acid which inturn reacts with the boiler metal to produce their metallic bicarbonate, and the cycle is repeated, resulting in severe corrosion.

$$2HCO_3^- \rightarrow H_2O + CO_3^{2-} + CO_2 \uparrow$$
$$CO_3^{2-} + H_2O \rightarrow CO_2 + 2OH^-$$
$$CO_2 + H_2O \rightarrow H_2CO_3$$
$$Fe + H_2CO_3 \rightarrow Fe(HCO_3)_2 + H_2$$
$$Fe(HCO_3)_2 + 2H_2O \xrightarrow{heat} Fe(OH)_2 + 2H_2O + 2CO_2$$

Carbon dioxide can be chemically removed by adding lime, soda ash or caustic- soda to form soluble bicarbonates or relatively insoluble carbonates. Another improved method is passing water through a bed of granular calcium carbonate. The calcium carbonate reacts with the carbon dioxide to form calcium bicarbonate as shown below.

$$CO_2 + H_2O + CaCO_3 \rightarrow Ca(HCO_3)_2$$

The major disadvantage with the chemical treatment is the production of additional bicarbonate / carbonate. This disadvantage is eliminated in the another method of removal of carbon dioxide namely, 'Deaeration', commonly employed at present.

Apart from corrosion, carbon dioxide does not exert any direct effect on industrial waters. It is indirectly controlled by the parameters such as pH, alkalinity and acidity. However, the presence of free carbon dioxide above 10 mg/l is undesirable for the manufacture of various types of papers.

CHAPTER – 8
CHLORIDE

Chloride is the common anion found in water and sewage. The concentration of chloride in natural waters varies from a few milligrams to several thousand milligrams per litre. (Sea water contains about 19,000 mg/l chloride). Higher concentration of chloride may be due to the contamination by sea water, brines, sewage or industrial effluents such as those from tanneries, textile processing units, paper mills, galvanizing plants, water softening plants and petroleum refineries. Therefore any increase in chloride concentration than the 'normal' indicates pollution.

Chloride produces a salty taste to water, but the salty-taste is variable and dependent on the chemical composition of the water. Reasonable amounts of chloride are tolerable in many potable water supplies and in industrial waters.

Chlorides in drinking water are generally not harmful to human beings. Higher concentrations however, may affect some persons who already suffer from disease of heart or kidneys. Generally, it is the cation (calcium, magnesium, sodium or potassium) associated with the chloride that produces a harmful effect.

Regarding irrigation waters, chloride is the most troublesome anion. Chlorides are generally more toxic than sulfates to most plants.

Chlorides in concentrations less than 50mg/l has no detrimental effects on most industrial uses. However waters containing higher concentrations are used rather widely. However, high concentrations exert deleterious effect on metallic pipes and structures. High chloride waters promote corrosion of metal surfaces. Waters containing excess amounts of chloride exert significant effect on the rate of corrosion. Chlorides though are not directly involved in corrosion, accelerate the rate of corrosion of steel, stainless steel alloys, and aluminium. Hence, the chloride concentration should be carefully controlled in systems where these materials are used.

The presence of even small amount of chloride as calcium chloride and magnesium chloride are highly undesirable due to their severe corroding effect on steam boilers. Besides, chlorides are also undesirable in process waters.

The recommended threshold concentrations for chloride ion and some chloride compounds are tabulated below.

TABLE - 8.1

RECOMMENDED THRESHOLD CONCENTRATIONS OF CHLORIDE AND CHLORIDE COMPOUNDS, IN mg/l*

Industry	Chloride	Calcium chloride	Magnesium chloride	Sodium chloride
Brewing	60 - 100	100 - 200	50 - 200	275 - 500
Carbonated Beverages	250	–	–	–
Dairy	30	–	–	–
Food Canning and Freezing	–	–	–	1000 - 1500
Food Equipment Washing	250	–	–	–
Paper making				
Ground wood pulp	75	–	–	–
Soda pulp	75	–	–	–
Kraft pulp	200	–	–	–
Steel manufacture	175	–	–	–
Sugar Making	20	–	–	–
Textile processes	100	–	–	–

Chloride concentrates in boiler water and evaporative cooling waters in an exact ratio to the amount of evaporation. Thus chloride determination of make up water and boiler water (or cooling water) is useful in estimating the amount of concentration that has occurred.

* Reproduced with the kind permission of the Resources Agency of California, State Water Resources Control Board, California from their publication "WATER QUALITY CRITERIA", 1963, edited by J.E. Mckee and H.W. Wolf.

CHAPTER - 9

CHLORINE

Chlorine does not occur naturally in water. Surface and ground waters may contain odor and taste producing substances (mostly organic in nature) the presence of which make them unsuitable for drinking and industrial uses. They are destroyed by chlorination converting them to innocuous compounds. Many industrial waters are chlorinated* to control bacteria and other slime–producing organisms. Potable waters are regularly chlorinated to render them bacteriologically safe. Hence the presence of chlorine is common in potable waters and industrial waters. Chlorine may also enter in to some surface waters through the discharge of industrial effluents containing chlorine (e.g. bleaching effluents from textile mills and paper mill effluents).

It is reported that in pure water, chlorine concentrations as high as 2.0 mg/l have not been found to cause objectionable tastes and odors. However the formation of undesirable tastes and odors are pronounced in the presence of even very small concentrations of organic matter such as phenol.

As stated earlier, many industrial waters are chlorinated to control bacteria and other organisms. But the chlorine concentration remaining is a cause of concern. The residual chlorine in the water used for food canning, may cause tastes in the canned foods or frozen products or it may corrode cans. For the production of high grade paper, the limiting concentration of chlorine is 0.3 mg/l and 0.8 mg/l for lower grade papers. It is also reported that traces of chlorine in water used for metal-plating baths will affect the smoothness or brightness of deposits.

In cooling towers, continuous contact with free chlorine may result in the deterioration of lumber. Further, large concentrations of chlorine can accelerate corrosion of metals particularly when oxygen is present.

---

* Industrial waters may contain bacteria (both pathogenic and non-pathogenic), algae, slimes, iron bacteria, manganese bacteria, sulfur bacteria and a variety of microscopic and macroscopic organisms. Besides, they may contain organic growths that may cause bad tastes and odors, and may also clog piping. Chlorination is the most effective method for the control of these organisms.

CHAPTER – 10

CHROMIUM

Chromium is not a natural constituent of water. It enters to a water course through the discharge of industrial effluents containing chromium. Electroplating wastes contain a high concentration of chromium. Chromium compounds are used extensively in cooling water systems as corrosion inhibitors. When such waters are discharged, chromium gains access to water. This apart, chromium compounds are used in high amounts in metal pickling, aluminium anodizing, leather tanning and in the manufacture of paints, dyes, paper, explosives and ceramics. Discharge of these wastes will also contribute chromium to water.

Chromium exists in two forms namely, hexavalent and trivalent. Chromates and dichromates are hexavalent chromium compounds. The toxic effects of chromium are attributed to hexavalent form only. The salts of trivalent chromium are physiologically not harmful.

Chromium when present in high concentration imparts color to receiving waters. Chromium is highly undesirable in a drinking water due to its suspected carcinogenic properties. Large doses of hexavalent chromium are reported to be corrosive in the intestinal tract.

If chromium enters into boilers, chromate will be reduced to trivalent chromium and gets precipitated as chromic hydroxide.

CHAPTER - 11
COLOR

Color is a common constituent of natural waters and is caused by humus materials (which is produced by the decay of vegetation), peat, tannin and metallic substances such as iron and manganese and chromium compounds. The products of corrosion and weeds and protozoa also contribute to color. Industrial waste waters also contribute to color.

Substances causing color normally exist in colloidal form. Hence mere filtration will not remove the true color of water. Coagulation can only remove these particles.

Many industrial processes need virtually colorless waters only. Determination of color is important with regard to industrial waters. The determination of color is a rapid one and is useful in detecting a change in the character of the water. When waters from the source as from river, are being regularly examined, the variations in color often serve as indices of quality. Generally yellow color indicates the presence of appreciable amounts of organic matter and sometimes chromium. The presence of peaty matter is indicated by the characteristic red brown color. The yellowish - red indicates the presence of iron. Any unusual color would indicate the necessity for a special examination to ascertain the cause. Color determination is also useful to check the functioning of filter beds and to determine the rate of filtration.

CHAPTER - 12
COPPER

Copper is rarely found in natural waters and when the amount exceeds 0.5 mg/l, it is attributable to the corrosive action of the water on copper and brass tubings, to industrial effluents or frequently to the use of copper compounds in controlling the growth of algae and other undesirable planktons in reservoirs. Ammonia when present in concentrations more than 10mg/l, along with dissolved oxygen usually results in fairly significant copper dissolution. Carbon dioxide and sodium bicarbonate present in water also corrode copper and copper alloys. Hence determination of copper in industrial water is necessary to monitor corrosion of copper, brass and other copper alloys.

Copper has the tendency to attack aluminium. Even very small amounts of copper will attack aluminium, particularly when present in hard water. Copper can be deposited in metallic form in pipes and boiler tubes with the dissolution of iron.

Presence of copper renders water disagreeable to taste. (Threshold concentrations for taste have been generally reported in the range of 1.0 - 2.0 mg/l of copper). In food processing and food preserving industries, copper causes undesirable color reactions, forming tannates and sulfides. In dairies copper in water acts as a catalyst for the oxidation of fatty matter and for the hydrolysis of milk constituents resulting in fatty, oily and fishy tastes in the products. Traces of copper in metal plating baths affect the smoothness and brightness of the metallic deposits. Copper is especially undesirable in the woollen industry for it is absorbed readily on wool. In the bleaching of wool, copper catalyzes the reaction with hydrogen peroxide to the extent that holes are formed in the material.

CHAPTER – 13

CYANIDE

Cyanide does not occur naturally in water. Cyanide if any detected in water is due to contamination by industrial effluents. Cyanides are likely to occur in dangerous concentration in waters contaminated with waste waters from electroplating and gold mining, from coke ovens and from the scrubbing of gases at steel plants. The chief reason for concern with cyanide is its high toxicity.

CHAPTER - 14

FLUORIDE

Fluoride occurs in almost all natural waters. Normally surface waters do not contain high amounts of fluoride unless polluted by industrial wastes such as from fertilizer industry. Fluorides may occur in detrimental concentrations in ground waters.

Fluorides when present in concentration range of 1.0 - 1.5 mg/l is an effective agent in preventing dental caries. Above this amount, it may cause dental and skeletal fluorosis. (Such waters have to be defluoridated to reduce the fluoride concentration to acceptable levels (0.5 - 1.5 mg/l) for drinking).

Fluoride has little significance in industrial waters, excepting food and beverage industries. The maximum allowable limit for fluoride in these waters is 1.5 mg/l.

CHAPTER - 15
HARDNESS (TOTAL)

Calcium and magnesium are the principal cations causing hardness. Hardness is deemed to be capacity of water for reducing and destroying the lather of soap. Hardness in water is due to the natural accumulation of salts from contact with soil and geological formations or it may enter from direct pollution by industrial effluents. Though calcium and magnesium are the principal cations causing hardness, iron, aluminінum, manganese, strontium and zinc also cause hardness but to a relatively little extent or to a negligible amount. The term “Total Hardness” thus indicates the concentration of calcium and magnesium ions only. However if present in significant amounts, the other metallic ions should also be included. The total hardness is expressed in terms of calcium carbonate.

The hardness of water varies considerably from place to place depending on the geological formations. Generally surface waters are softer (containing much less quantities of calcium and magnesium) than ground waters.

Hardness has traditionally been classified as ‘Temporary Hardness’ and ‘Permanent Hardness’. The portion of hardness that disappears on prolonged boiling is referred as temporary hardness and is mainly due to the bicarbonates of calcium and magnesium which are precipitated as normal carbonates by the loss of carbon dioxide on heating. The hardness that remains after boiling is known as permanent hardness and is due to the sulfates, chlorides and nitrates of calcium and magnesium. However these terms are obsolete and the more precise terms are ‘Carbonate Hardness’ and ‘Non-carbonate Hardness’. The carbonate hardness is due to the bicarbonates of calcium and magnesium while non-carbonate hardness is due to sulfates, chlorides and nitrates of calcium and magnesium. When the hardness is numerically greater than the sum of bicarbonate and carbonate alkalinities, the amount equal to the alkalinity is carbonate hardness and the excess amount is non-carbonate hardness. If the hardness is equal to or less than of bicarbonate and carbonate alkalinities, all of the hardness is carbonate hardness and there is no non-carbonate hardness.

The major scale forming constituents in water are calcium and magnesium ions only. They form heavy voluminous deposits ranging from hard scales to soft precipitates. Practically all boiler water treatment methods aim in reducing the calcium and magnesium content of water.

Carbonate hardness (bicarbonate / carbonates of calcium and magnesium) is generally not objectionable for industrial uses. If it is excessive, it can be removed or reduced rather simply at low cost. As for steam generation is concerned when the water containing carbonate hardness is boiled, carbon dioxide is rapidly given off and calcium carbonated is deposited. If the amount of bicarbonate does not exceed 100 mg/l, the insoluble carbonate formed does not form scale, but comes out of solution as a sludge which can be removed by blowing down. The same applies to magnesium carbonate also.

When the amount of bicarbonate exceeds 100 mg/l, calcium bicarbonate and magnesium bicarbonate breakdown to form scales. The calcium scale consists of calcium carbonate and magnesium deposits as magnesium hydroxide. Magnesium bicarbonate first decomposes to magnesium carbonate which in turn reacts with water to form magnesium hydroxide. The magnesium hydroxide loses part of its water content by baking on the tubes and appear partly as magnesium oxide. The reactions are illustrated in the following equations.

$$Ca(HCO_3)_2 \rightarrow CaCO_3 + H_2O + CO_2 \uparrow$$
$$Mg(HCO_3)_2 \rightarrow MgCO_3 + H_2O + CO_2 \uparrow$$
$$MgCO_3 + H_2O \rightarrow Mg(OH)_2 + CO_2 \uparrow$$

The carbon dioxide formed during these reactions leaves boiler with the steam.

Calcium carbonate occurs either as a soft deposit or as a hard scale. It is a major constituent of deposits in preboiler systems. Magnesium hydroxide and magnesium silicate generally form soft deposits and sludge. These interfere with heat transfer and reduce efficiency. These deposits have to be removed periodically. Otherwise tube failures may occur frequently.

Non carbonate hardness (sulfates, chlorides and nitrates of calcium and magnesium) is highly undesirable for industrial purposes. It is very difficult to remove and cost for reduction /

removal is high. The non-carbonate hardness is especially troublesome when the anion is sulfate. When waters containing calcium sulfate is used in boilers, it does not precipitate until the water is concentrated. It then crystallizes out as a hard and adherent scale on the surface of the boilers which is very difficult to remove even mechanically. As the thickness of the scale increases, the boiler plates may become seriously over heated and distorted. Magnesium sulfate is more soluble and does not form scale. However if the concentrated solution is not blown down frequently, it may also cause serious trouble.

The chlorides of calcium and magnesium are also highly objectionable as the concentrated solutions when boiled, give off hydrochloric acid in the steam causing severe corrosion (pitting).

Besides sulfates and chlorides of calcium and magnesium, nitrates are also undesirable in feed waters since they are reduced to ammonia by the iron of the boiler plates and is given off in the steam.

Silica combines with calcium and magnesium salts to form a hard mass and produce the hardest of scales. The occurrence of such silica scales on turbine blades is not uncommon (refer Chapter - 34 "SILICA"). Not only for boiler purposes, but that waters should necessarily be treated for the removal of calcium and magnesium when the water is intended for cooling. These salts form incrustation and deposits scale in open cooling system. (During evaporative cooling, the concentration of dissolved solids increases, and the solubility product exceed with the resultant formation of scale or sludge of calcium carbonate). In the open recirculation system, calcium sulfate also precipitates which may deposit over the heat exchange surfaces. To avoid precipitation of calcium sulfate, the bleed off has to be properly adjusted.

Hard waters are generally undesirable for dyeing and textile industries. Calcium and magnesium ions precipitate certain dyestuffs, producing streaky or spotty dyed goods, or they may act as mordants, causing greater dyestuff absorption and deeper shades.

Because of excessive soap consumption in laundries only soft water is preferred.

In the canning of fruits and vegetables, salts of calcium and magnesium may combine with pectous substances forming insoluble pectides that toughen the product. Hard water toughens peas and some varieties of beans also when the hardness exceeds 200 mg/l.

In bakeries, very hard waters are reported to retard fermentation significantly.

Hardness is desirable in certain instances however. For example, in dyeing, a minimum amount of magnesium is required. In brewing, waters containing calcium sulfate is highly preferable. In baking, medium-hard water strengthens gluten causing it to retain carbon dioxide better and produces a finer grained structure. Soft waters soften the gluten thus producing a sticky dough and a soggy bread.

Very soft waters having hardness less than 30 mg/l are likely to be corrosive; waters with very low hardness (less than 10 mg/l) attack concrete and metallic structures. Usually such waters are treated with lime to increase the hardness.

The levels of hardness tolerable or desirable in water for various industrial purposes are summarized in the following table.

Table – 15.1

RANGE OF RECOMMENDED VALUES FOR HARDNESS IN INDUSTRIAL WATER SUPPLIES*

Industry and process	Limiting or recommended Value in mg/l
Boiler Feed Water	
at 0 - 150 psi	80
at 150 - 250 psi	40
at 250 - 400 psi	12
over 400 psi	2
Brewing	200 - 300
Carbonated beverages	200 - 250
Cooling	50
Food Canning and Freezing	
General	50 - 85
Legumes	25 - 75
Fruits and vegetables	100 - 200
Peas	200 - 400
Food Equipment Washing	10
Food Processing (General)	10 - 250
Laundering	0 - 50
Pulp and Paper Making	
Ground wood pulp	200
Soda pulp	100
Kraft pulp, bleached	100
Kraft pulp, unbleached	200
Fine paper pulp	100
Rayon	
Pulp production	8
Cloth manufacture	55
Steel manufacturing	50
Synthetic Rubber	50
Tanning	
Beam house	513
Tan house	50 - 135
Textile manufacture	0 - 50

* Reproduced with the kind permission of the Resources Agency of California, State Water Resources Control Board, California from their publication "WATER QUALITY CRITERIA", 1963, edited by J.E. Mckee and H.W. Wolf.

CHAPTER - 16
HYDRAZINE

Hydrazine does not occur naturally in water. It is added to boiler feed water and boiler water to remove the residuals of oxygen. Hydrazine is the best-known oxygen scavenger and it reacts with oxygen to form water and nitrogen gas. Its greatest advantage is that the reaction products are harmless and do not add to the dissolved solids in the boiler. However large excess are harmful as they decompose to yield ammonia to the steam, which is highly detrimental because when the steam condensate contacts the parts made up of copper alloys, severe corrosion takes place, in the presence of oxygen.

CHAPTER - 17

HYDROGEN SULFIDE

Hydrogen sulfide is a toxic gas and is notorious for its offensive odor. It is highly soluble in water. It is occasionally found in surface waters resulting from pollution or anaerobic decomposition of organic matter. In some ground waters also it is found as a result of reduction of sulfate (by sulfate reducing bacteria). Waters which contain sulfides are commonly known as "sulfur waters". They are corrosive in nature besides emanating an offensive odor.

Hydrogen sulfide is the most undesirable substance in industrial waters, especially those used for foods and beverages. Besides imparting the disagreeable odor to the product, it also damages the equipment. In relatively higher concentrations it will attack normally resistant alloys such as stainless steel and brass and cause rapid deterioration. The maximum prescribed limits for hydrogen sulfide in industrial waters are tabulated below.

TABLE – 17.1
RECOMMENDED THRESHOLD VALUES FOR HYDROGEN SULFIDE IN INDUSTRIAL WATERS*

Industrial use	Recommended threshold values in mg/l
Baking	0.2
Brewing	0.2
Carbonated beverages	0.0 - 0.2
Confectionary	0.2
Cooling water	5.0
Food canning and freezing	1.0

Hydrogen sulfide itself is a corrosive gas and affects boiler metal. Hydrogen- sulfide is corrosive because it ionizes to form a weak acid as shown below.

$$H_2S \rightarrow H^+ + HS^-$$

* Reproduced with the kind permission of the Resources Agency of California, State Water Resources Control Board, California from their publication "WATER QUALITY CRITERIA", 1963, edited by J.E. Mckee and H.W. Wolf.

Whenever H_2S is present in water, iron sulfide is formed. These deposits are cathodic to base metal and hence followed by severe pitting beneath the iron sulfide deposit. When oxygen intrudes into sulfide system the rate of corrosion becomes uncontrollable. Carbon dioxide mixed with H_2S is also much more aggressive than either of the gases alone.

Though the feed water does not contain any sulfide it should be produced insitu in the boiler. Under an operating temperature of 270°C and above, sodium-sulfite treated water breaks down to sodium sulfide and sodium tetrathionate. The sodium- sulfide is then hydrolyzed to hydrogen sulfide and passes in to the steam and cause severe corrosion. The lurking danger with hydrogen sulfide is that under certain conditions it may even be oxidized to sulfuric acid leading to severe corrosion of pipes and metallic structures carrying and holding water to a significant extent. Besides, it is capable of precipitating many cooling water treatment chemicals, causing unwanted deposits. Because of these objectionable characteristics, removal of hydrogen sulfide from process waters and waters intended for cooling and boiler feed is an essential step in water treatment program. Aeration and chlorination are some of the common methods employed. Addition of permanganate is also practised to oxidize sulfide. There is a danger in aeration in that hydrogen sulfide is some times converted to colloidal sulfur. Chlorination is found to be more effective and a low cost method. However care should be exercised in selecting the technic and chemical as the aqueous system contains a variety of species, some of which may take part in the oxidation reaction leading to some other problems.

CHAPTER - 18
IRON

Iron is one of the important elements, essentially required for the nutrition and healthy development of man, animals and plants. Though iron ranks next to aluminium in abundance of metals in the earth's crust, natural waters contain only minor amounts of iron. However, iron content of some deep wells and industrially contaminated streams may be very high.

Iron usually occurs in the ferrous form as the soluble colorless ferrous bicarbonate. Such waters are usually perfectly clean and colorless when first drawn. However on exposure to air, the water becomes turbid and discolored due to the oxidation of soluble ferrous bicarbonate to colloidal ferric hydroxide. When the quantity of iron is more, yellowish to reddish brown precipitate of ferric hydroxide is formed.

Iron in water is highly objectionable both for domestic and industrial uses owing to the production of discoloration, turbidity and taste. Ferruginous waters have a characteristic astringent (or bitter) taste. Iron combines with tannin (derived from plants) and the water acquires an inky color.

All industrial processes require water virtually free from iron (and manganese). The tolerance limit of 0.05 to 0.1 mg/l iron is prescribed by the Technical Association of the Pulp and Paper Industry, as iron produces stains and off-colors, and also causes loss in bleaching. In tanneries, iron produces spots and stains on leather. Iron bearing waters impart an off-white color to clothes when laundered. Both iron and manganese tend to precipitate as hydroxides and stain porcelain fixtures. In brewing, water with an iron content more than 0.1 mg/l is undesirable as it may impart an unpleasant taste and odor and a greenish tint to the beer.

In dairies also, water free from iron and manganese is required as these metals functions as catalysts for the oxidation of fatty matter and for the hydrolysis of milk constituents leading to fatty, oily or fishy tastes in the milk products.

When ferruginous waters are used for culinary purposes, iron causes brown-colored deposition on vegetables while washing and cooking.

For ice manufacturing, absence of iron and manganese is essential. They result in a colored and an unattractive ice cake.

As records textile processing (bleaching, dying, printing etc), the total absence of iron and manganese is desirable. The various effects caused by these metals are dealt exhaustively in Chapter - 90 "TEXTILE PROCESSING".

Besides their detrimental effects on industrial uses, iron and manganese are of serious concern in boiler waters and cooling waters.

Corrosion of metals in boiler and pre boiler systems release iron into water which cause serious problems. The released iron will be converted to iron oxide which circulate with the water and may not be removed by blow down. Such oxides form adherent coatings an the surface of the boiler and lead to tube failures. The iron may also combine wish sodium silicate (if present) to form a very hard and adherent deposit over the surfaces of high heat transfer.

Very serious fouling problems are attributed to cooling waters containing iron (dissolved or suspended). Fouling occurs due to deposition of iron oxide or by the reaction of iron with treatment chemicals such as chromate and phosphate.

Iron and manganese are the most troublesome constituent and they are the causative agents for serious fouling of ion-exchange resins during water softening. The precipitates of iron and manganese would cause clogging problems and would coat the surface of the resin resulting in a heavy loss of exchange capacity.

Even traces of iron and manganese in water lead to the accumulation of appreciable deposits in distribution mains and are very troublesome. Under such conditions, growths of iron bacteria often develop in the distribution system causing difficulties such as blocking of mains and pipes, thus reducing their carrying capacity. Any attempt to slough off these accumulations makes the water discolored, turbid and unpalatable. This apart, the deposited iron and manganese precipitates are frequently resuspended by increased flow rates imparting high turbidity to water. Sometimes iron sulfide and/ or hydrogen sulfide may also be produced, which emits an offensive odor when water drawn from the taps often misleading that the water is contaminated with sewage. The so called "red water trouble" is due to severe corrosion of iron in distribution systems. The iron subsequently is precipitated as ferric hydroxide which imparts a rusty color and turbidity to the water.

Similar to iron, manganese also causes "black water trouble". Due to corrosion, manganese is released which forms black precipitate that may accumulate in distribution system and released occasionally.

The ranges of recommended threshold values for iron in the process and cooling waters are tabulated below.

TABLE – 18.1
RANGE OF RECOMMENDED THRESHOLD VALUES FOR IRON IN INDUSTRIAL WATERS, IN mg/l*

Industrial Use	Range of Recommended Threshold Values, in mg/l
Baking	0.2
Brewing	0.1 – 1.0
Carbonated beverages	0.1 – 0.2
Cooling water	0.5
Confectionary	0.2
Electro plating	Traces
Food Canning and Freezing	0.2
Food Equipment Washing	0.2
Food Processing, general	0.2
Laundering	0.2 – 1.0
Oil well flooding	0.1
Photographic processes	0.1
Pulp and Paper making	
Ground wood pulp	0.3
Soda pulp	0.1
Kraft pulp, bleached	0.2
Kraft pulp, unbleached	1.0
Fine paper pulp	0.1
Rayon Manufacture	0.0 – 0.5
Sugar making	0.1
Tanning processes	0.1 – 2.0
Textile manufacture	0.1 – 1.0

* Reproduced with the kind permission of the Resources Agency of California, State Water Resources Control Board, California from their publication "WATER QUALITY CRITERIA", 1963, edited by J.E. Mckee and H.W. Wolf.

CHAPTER - 19

LEAD

Lead is not practically occurring in any natural water. If lead is present in a water supply, it may be due to the corrosion reactions and waste water contamination. Mine waters, electroplating wastes and lead paint wastes contribute lead to raw water supplies. Waters in contact with lead piping or lead bearing jointing compounds, usually contain lead. Since lead is a cumulative poison, its determination in potable waters is important. For industrial uses, lead is not usually a significant problem. However, it is reported that traces of lead in metal–plating baths will affect the smoothness and brightness of deposits.

CHAPTER - 20

MAGNESIUM

(See also, Chapter - 15 "HARDNESS (TOTAL)

Magnesium salts occur in significant concentration in natural waters. Sea waters and estuary waters contain high amounts of magnesium. (Magnesium content of sea water is about 5200 mg/l as $CaCO_3$).

Magnesium is an essential element for human beings. Magnesium is relatively non-toxic to man. However, higher concentrations cause unpleasant tastes to water. At high concentrations, magnesium salts have a laxative effect particularly when present as magnesium sulfate.

Magnesium in industrial water is undesirable since it is the major scale forming cation. Magnesium salts, such as magnesium silicate and magnesium hydroxide, generally form soft deposits and sludges. At the temperature prevailing in steam boiler, both calcium bicarbonate and magnesium bicarbonate breakdown to form scale. Calcium bicarbonate decomposes to calcium carbonate while magnesium bicarbonate decomposes to magnesium carbonate first which immediately is converted to magnesium hydroxide.

$$Ca(HCO_3)_2 \rightarrow CaCO_3 + H_2O + CO_2 \uparrow$$
$$Mg(HCO_3)_2 \rightarrow MgCO_3 + H_2O + CO_2 \uparrow$$
$$MgCO_3 + H_2O \rightarrow Mg(OH)_2 + CO_2 \uparrow$$

These deposits interfere seriously with heat transfer and reduce efficiency. If such deposits are not removed periodically, tube failures may occur frequently. The magnesium bicarbonate content of natural waters are normally lower than calcium bicarbonate content and hence in once-through cooling systems, magnesium plays only a minor role in scale formation.

Magnesium chloride and magnesium sulfate are noted for their corrosiveness. Magnesium chloride is especially corrosive and the action seems to be cyclic in nature. The magnesium chloride first decomposes to magnesium hydroxide and hydrochloric acid. The formed hydrochloric acid attacks the boiler metal which results in the formation of ferrous chloride. This ferrous chloride reacts with the magnesium hydroxide formed initially to form ferrous hydroxide and magnesium chloride and the cycle is repeated. (Dissolved oxygen from partially deaerated or undeaerated water plays a vital role in such attacks). If sodium alkalinity is present, magnesium chloride/ magnesium sulfate reacts to form magnesium hydroxide so that this type of attack would not occur.

Magnesium in process water is not an objectionable element excepting for certain processes. The threshold concentration of magnesium for various industrial uses are tabulated below.

TABLE - 20.1
RECOMMENDED THRESHOLD CONCENTRATION OF MAGNESIUM IN INDUSTRIAL WATERS*

Process	Recommended limits in mg/l for			
	Magnesium	Magnesium sulfate	Magnesium chloride	Magnesium bicarbonate
Brewing	30	100 - 200	100 - 200	–
Ice manufacture	–	130 - 300	171 - 300	50
Soda pulp	12	–	–	–
Sugar making	10	–	–	–
Textile manufacture	5	–	–	–

Magnesium infact is desirable in some process waters. Like calcium, magnesium in small amounts is beneficial in the mash water for pale beer.

* Reproduced with the kind permission of the Resources Agency of California, State Water Resources Control Board, California from their publication "WATER QUALITY CRITERIA", 1963, edited by J.E. Mckee and H.W. Wolf.

CHAPTER - 21

MANGANESE

(See also, Chapter - 18 "IRON")

Manganese occurs in soils and rocks as manganese dioxide and can be dissolved in natural waters by the action of anaerobic bacteria. Under reducing conditions, manganese can be leached from the soil and occur in considerable concentrations in ground water.

Manganese is undesirable in domestic water supplies as it causes unpleasant tastes, deposits on food during cooking, stains and discolors laundry and plumbing fixtures. Even very small amounts of manganese (0.2 – 0.3 mg/l) may form heavy incrustations in piping. It is also reported that even smaller amounts may form noticeable black deposits. It also fosters the growth of certain micro-organisms such as Crenothrix and Gallionella in reservoirs, filters and distribution systems.

Since manganese exhibits objectionable characteristics similar to those of iron, excessive concentration of manganese is undesirable in most of the industrial water supplies. Particularly, industrial processes such as paper manufacturing, brewing, dyeing and ice manufacturing and textiles industries need water free from manganese. In cooling systems, manganese causes corrosion reactions in copper alloy tubing. Manganese in boiler water may also accelerate corrosion.

The following tabulation summarizes the recommendations as to maximum permissible concentrations of manganese in industrial waters.

TABLE - 21.1

MAXIMUM PERMISSIBLE CONCENTRATION OF MANGANESE IN INDUSTRIAL WATERS*

Industrial Use	Maximum Permissible Concentration	
	Manganese in mg/l	Iron + Manganese in mg/l
Air-conditioning	0.5	0.5
Baking	0.2	0.2
Brewing (light & dark)	0.1	0.1
Canning	0.2	0.2
Carbonated Beverages	0.2	0.2
Confectionary	0.2	0.2
Cooling water	0.2	0.2
Dyeing	0	0
Food processing	0.2	0.2
Ice	0.2	0.2
Milk industry	0.03 - 0.1	–
Paper and pulp	0	0
Ground wood	0.5	1.0
Kraft pulp	0.1	0.2
Soda and Sulfate	0.05	0.1
High grade paper	0.05	0.1
Fine paper	0.05	0.1
Kraft paper		
Bleached	0.1	–
Unbleached	0.5	–
Photography	0	0
Plastics (clear)	0.02	0.02
Rayon and viscose		
Pulp production	0.03	0.05
Manufacture	0	0
Tanning	0.2	0.2
Textiles, General	0.25	0.25
Wool scouring	1.0	1.0
Bandages	0.2	0.2

* Reproduced with the kind permission of the Resources Agency of California, State Water Resources Control Board, California from their publication "WATER QUALITY CRITERIA", 1963, edited by J.E. Mckee and H.W. Wolf.

CHAPTER - 22
MERCURY

The occurrence of elemental mercury in water is rare. Mercury compounds are widely employed in the manufacture of disinfectants, detonators, pigments and medical products. Discharge of these wastes may contribute mercury to the receiving waters. Mercury can also enter into a water course through agricultural pesticides, herbicides, and fungicides. Among the compounds of mercury found in water, methyl mercury and phenyl mercuric acetate are the most common.

Mercury and mercuric compounds are considered to be highly toxic to human beings, animals and aquatic life. Prolonged mercury ingestion can cause kidney damage, brain damage and loss of muscle control. Hence the stringent limit of 0.001 mg/l is imposed on waters used for drinking, food preparation, food canning and ice manufacture etc.,

CHAPTER - 23

NICKEL

Nickel does not occur in natural water. Metal plating waste waters when discharged to water courses, contribute nickel. Increased use of stainless steel and other nickel containing alloys in water systems also contribute nickel. The determination of nickel is of much importance in water chemistry for high pressure boilers because of the following reasons:

i. The presence of nickel indicates corrosion of nickel alloys.

ii. It behaves similarly as copper and may be precipitated either as the oxide or as the metallic nickel.

Nickel is considered relatively non-toxic to man. However it is reported to be injurious to several plant species. It is also reported to cause adverse effects on biological sewage treatment and on sludge digestion.

CHAPTER - 24

NITRATE

Nitrates are the end products of the aerobic stabilization of organic nitrogen and occur generally in trace quantities in surface water supplies but may attain high levels in some ground waters. A nitrate nitrogen concentration of 70 mg/l in ground waters is not uncommon. Application of fertilizers to lands and leachings from cesspool contributes nitrate to ground waters. Waste waters from chemical, fertilizer manufacturing plants are also important contributors to nitrate. Besides, the strata through which the water percolates is also a source of significant amount of nitrate.

Nitrate is an important parameter with regard to potable waters. It has been reported that nitrate concentration above 20mg/l may cause 'methemoglobinemia' in infants, a disease characterized by blood changes. The maximum limit for nitrate in drinking water as per WHO drinking water standards is 50 mg/l nitrate.

Nitrates are reported to be injurious to dyeing of wool and silk fabrics. They are also harmful to fermentation processes and cause disagreeable tastes to beer. For brewing, the nitrate concentration in water should not exceed 15 – 20 mg/l. In brewing industry, during fermentation, nitrates are partially reduced to nitrites which behave as poison to yeast. When present in high concentrations, nitrates impart a bad taste to beer.

On the other hand, nitrates are used in water treatment (in corrosion-inhibitor compounds) for cooling system. Nitrates alone or in conjunction with certain organic compounds are also effective agents in preventing caustic embrittlement (inter-granular corrosion) in low pressure boilers.

CHAPTER - 25

NITRITE

In water, nitrites are generally formed by the action of bacteria upon ammonia and organic nitrogen. Nitrite forming bacteria convert ammonia to nitrite under aerobic conditions. Sometime nitrites are also produced by the reduction of nitrates under anaerobic conditions. Hence the presence of nitrite in water indicates organic pollution (i.e) more precisely it indicates that the nitrogenous organic matter is undergoing oxidation or nitrification and that the process is not complete. Owing to the fact that nitrites are readily oxidized to nitrates, they are seldom present in significant concentrations in surface or other natural waters.

Nitrite compounds are added as corrosion inhibitors to cooling waters and to industrial process waters. In food industry, nitrite compounds are used as preservatives. (Hence nitrite may also enter into water supply through the discharge of these waste waters.)

Waters containing nitrite are undesirable for industrial purposes, particularly for brewing and dyeing of wool and silk fabrics.

CHAPTER – 26

NITROGEN (AMMONIA & ORGANIC)

Nitrogen occurs abundantly in nature, constituting about 78% by volume of air. In natural or polluted waters nitrogen may occur in many forms such as nitrate, nitrite, ammonia and organic nitrogen. Nitrate nitrogen and nitrite nitrogen are discussed as separate Chapters previously.

Ammonia generally arises from the aerobic or anaerobic decomposition of nitrogenous organic matter. Ammonia as such may not be present as free ammonia gas but it would be present in the form of salts such as ammonium chloride, ammonium carbonate etc. Since ammonia is derived from the decomposition of sewage and from agricultural run off, the presence of ammonia in water is often regarded as indicative of fresh pollution by nitrogenous wastes. The presence of more than traces of ammoniacal nitrogen in water is undesirable. Waters with high ammoniacal nitrogen content have a high chlorine demand and require much longer contact periods for satisfactory sterilization.

Organic nitrogen is the organically bound nitrogen in nitrogenous organic matter such as proteins, amino acids, and organic bases. Hence an increase in the organic nitrogen content of a water may indicate significant organic pollution.

The term 'Total Kjeldahl Nitrogen' is used to denote the sum of ammoniacal nitrogen and organic nitrogen.

CHAPTER - 27

OIL AND GREASE

The presence of even very low concentrations of oil and greasy matter in raw water sources is undesirable due to the complications produced by them in the treatment processes such as coagulation, and sedimentation. If any of the free or emulsified oil reaches the sand filters or ion exchange beds, it will coat the grains and decrease the effectiveness of filtration and also interfere with the back washing.

The presence of oil in boiler feed water may cause foaming, priming, overheating of tubes resulting in bulging or even bursting. It is to be mentioned that cooling waters should also be free from oil otherwise biological slimes may occur. Most of the industrial process require only oil free water, because of the possibility of production of unwanted tastes and odors in the manufactured products due to the oil. This is especially true for brewing, carbonated beverages, bakeries, food canning and freezing, ice manufacture, sugar production and food-equipment washing.

Oil is also detrimental in water used for paper making in that it may adhere to the stock, form slick spots on the paper, or promote the growth of undesirable organisms.

CHAPTER - 28

ORGANIC MATTER

There is an astounding variety of organic compounds present in water. They are contributed to water course by the discharge of industrial effluents and sewage. Agricultural activities and decomposition of vegetable and animal matter and soil wash also adds organic matter to water. Organic materials do not occur as single component. Usually a number of compounds are present.

Color of a water usually results due to the presence of organic matter. Organic matter is also the chief source of taste and odor. Organic matter in water is highly objectionable to almost all industrial processes as it gives rise to many problems. In food industry, many forms of organic matter impart an unpleasant taste and odor to the food products. They are reported to cause putrefaction. In textile processing, they discolor the fibres and fabrics and interfere in dyeing. Animal organic matter is reported to cause grease spots on fabric and paper which will not take up the dye or ink. They also cause foaming in boilers.

Removal of organic matter from water is an essential step in water treatment to make it suitable for industrial purposes. Oxidation with chlorine, chlorine dioxide or ozone completely converts the objectionable organic matter to innocuous substances.

CHAPTER - 29

OXYGEN (DISSOLVED)

(See also, Chapter - 7 "CARBON DIOXIDE")

Oxygen is dissolved in most waters in varying concentrations. Dissolved oxygen in surface waters ranges from 3 to 12 mg/l. Polluted waters normally contain very low amount of dissolved oxygen (less than 3mg/l) and sometimes completely devoid of oxygen under extreme situations.

Solubility of oxygen depends on temperature, pressure and salinity of water. It is essential to the fish life and other aquatic organisms. However in industrial waters, dissolved oxygen is a nuisance as it induces corrosion reactions. A solution of oxygen in water is very corrosive to metals such as iron, steel, galvanized iron and brass, which are most widely used for making vessels for holding and conducting water. Low pH values accelerate the rate of this dissolved oxygen corrosion; high pH values tend to retard it.

Most of the industries need water virtually free of oxygen. Oxygen is considered as the chief culprit in boiler system corrosion. Even trace amounts of oxygen in boiler water, feed water and cooling water can cause corrosion. For boiler feed water, the limiting concentration of DO is 2.0 mg/l for boiler pressures up to 150 psi; 0.2 mg/l for pressures from 150 to 250 psi; and total absence for 250 psi and above.

Dissolved oxygen corrosion is often very severe because of air leaking into the systems at threaded joints, through faulty steam traps and packing glands at periods when the steam is off. Its corrosion action is much more rapid than that of carbon- dioxide. Due to extremely low alkalinity, the pH value of the condensate would be very low. Such low pH is further reduced by the presence of even small concentrations of carbon dioxide. Carbon dioxide itself is a corrosive agent and also accelerate the rate of dissolved oxygen corrosion. In combination they are more virulent than either alone, especially if the pH is lower than 6. It is also to be mentioned that corrosion takes place only if liquid water is present. Accordingly dry steam which contains oxygen is not corrosive, but the condensate formed from such steam is very corrosive. In condensate return lines it will be found that the greatest attack is at the places where liquid water is pocketed.

It is well known that the elevation of temperature greatly accelerates the rate of corrosion. Theoretically the corrosion rate would be 500 times as fast at 90°C as at 0°C. (But at higher temperatures, the oxygen content diminishes to a significant extent and the rate of reaction is reduced considerably). In practice, it is observed that the greatest speed of corrosion seems to take place in the temperature range of 70° – 82°C.

To combat the dissolved oxygen-corrosion problem, several methods are practised. As regards boiler feed water, deaeration is noteworthy which is reported to reduce the oxygen content even below 0.005 mg/l. Feeding of oxygen scavenging chemicals (sodium - sulfite, hydrazine and other proprietary products) in to steam boilers remove any residual oxygen. In closed recirculation cooling systems and for a number of other uses, chromate is used as a corrosion inhibitor. The most widely employed method for protecting the tanks and pipeline is cathodic protection. (Detailed discussion of these methods is beyond the scope of this book. Excellent reference books are available on these methods and readers may refer such books).

CHAPTER - 30

PHENOLS

Phenols and phenolic compounds are not present in natural waters and are contributed by the effluents from oil refineries, gas works, coke plants and chemical plants. Phenols impart a characteristic objectionable odor (and tastes) to receiving waters. Phenols if present even in very low concentrations (0.001 mg/l) are able to produce tastes and odors in water, particularly during chlorination. Phenol is highly undesirable in many of the food and beverage industries due to the formation of such obnoxious tastes and odors. That is why the stringent limit of 0.001 mg/l is prescribed for phenols in water intended for drinking and food processing including ice manufacture. This limit is exclusively based on taste and odor rather than toxicological considerations.

The removal of phenolic tastes and odors from a water supply is a serious challenge to Water Engineers and Water Chemists. Activated carbon adsorption, super- chlorination, chlorine-ammonia treatment are some of the processes used for the removal of phenol from water.

CHAPTER - 31

PHOSPHATE

Phosphate may occur in surface or ground waters as a result of leaching from minerals or ores, from agricultural run off, or as a result of discharge of industrial wastes. Municipal sewage also contains phosphates (due to the use of synthetic detergents) to a significant concentration and when discharged to water courses contribute phosphates. Surface waters seldom contain higher concentrations of phosphate since they are utilized by plants whereas ground waters usually contain appreciable amounts of phosphates.

Phosphates and polyphosphates are the widely employed chemicals to prevent scale formation and to inhibit corrosion, particularly in boiler feed waters and cooling waters.

Phosphates are used almost universally as scale preventing chemicals in boilers operating above 400 psi. Any residual calcium entering a phosphate treated boiler water is precipitated as soft sludge of calcium phosphate and removed by blow down. Even though the feed water is entirely free from calcium and magnesium, phosphate excess is often maintained in the boiler water as a guard against unexpected contamination (or hardness pick up from condenser tube leakage).

Similarly phosphates are extensively used both in once–through and recirculating cooling water systems for scale prevention and corrosion inhibition. In cooling systems, a thin film of iron meta- phosphate is formed over the metallic surface which inhibits corrosion.

Despite the merits mentioned above, presence of phosphate is undesirable in certain instances as they foster algal growth which is detrimental. Modern synthetic detergents invariably contain sodium tripoly phosphate and tetrasodium pyrophosphate as builders and waters naturally contain these chemicals in increasing concentration. Polyphosphates interfere with the treatment processes such as coagulation, flocculation and lime soda treatment of water.

In certain industrial processes also, phosphates are undesirable. For example, in brewing, the phosphates make the wort more alkaline and affects the aroma of the beer.

CHAPTER - 32
pH VALUE

pH is a term used universally to express the intensity of the acid or alkaline condition of a solution. It is a measure of hydrogen ion concentration, or more precisely, the hydrogen ion activity.

pH is defined as the 'logarithm' (base 10) of the reciprocal of the hydrogen-ion concentration. Thus if $[H^+] = 10^{-6}$ moles per litre, then pH = 6. pH is an important factor in water chemistry, since it enters into the calculation of acidity and alkalinity and processes such as coagulation, disinfection, softening and corrosion control.

The pH of a raw water used for domestic purposes is important in that it affects taste, efficiency of chlorination and treatment processes such as coagulation. The killing power of chlorine diminishes with increasing pH values and consequently it is advantageous to apply chlorine to a water with a pH value of 7 or less. Knowledge of pH is essential in the selection of coagulants for water purification. For example, aluminium - sulfate is more effective at a pH 6.7, whereas ferrous sulfate coagulates well at a higher pH. Ferric sulfate is effective over a wide range.

pH in conjunction with total salinity, temperature, calcium content and total alkalinity is used to determine whether a water is corrosive in nature or having scale forming tendencies. Langlelier saturation index and Ryznar stability index based on the actual pH values and theoretical pH values are the widely employed parameters to assess the character of a water - either corrosive or having the tendency to form scale.

Maintaining the proper pH of water is very essential in industrial operation. The pH limitations and the desirable pH range for industrial waters are tabulated below.

TABLE – 32.1

RECOMMENDED THRESHOLD VALUES OF pH FOR WATERS USED IN VARIOUS INDUSTRIAL PROCESSES*

Industrial process	Recommended pH value Minimum	Range
Boiler feed water		
0 - 150psi	8.0	—
150 - 250psi	8.4	—
250 - 400psi	9.0	—
400 psi and over	9.6	—
Brewing	—	6.5 - 7.0
Confectionary	7.0	—
Food canning and freezing	7.5	—
Laundering	—	6.0 - 6.8
Oil-well flooding	7.0	—
Rayon manufacture	—	7.8 - 8.3
Steel making	—	6.8 - 7.0
Tannery operation	—	6.0 - 8.0

Some of the interesting observations made by various researchers are reported below:

1) For high-grade pulp and paper, Fleming etal† proposed that the pH of the water supply be 7.0 and for lower-grade products 6.7 - 7.3

2) Collins‡ states that distilled water is not satisfactory for making hard candies because it is likely to have a pH of 6.1 - 6.5 which favors inversion of sucrose and makes the candy sticky.

* Reproduced with the kind permission of the Resources Agency of California, State Water Resources Control Board, California from their publication "WATER QUALITY CRITERIA", 1963, edited by J.E. Mckee and H.W. Wolf.

† Fleming, Hopkins, Imhoff, Macdonald etal, "Industrial Water for Pulp, Paper and Paperboard Manufacture" TAPPI Monograph series No. 1 published by TAPPI, New York, 1942

‡ Collins W.D, "Water for Industrial Purposes", American City (July, Aug and Sept 1937).

CHAPTER - 33

POTASSIUM

Potassium ranks seventh among the elements in the order of abundance. However its concentration in most water is trivial. Since potassium is usually present in much smaller amounts and is so much like sodium in most properties, is usually grouped with sodium. Potassium is an essential nutritional element, but in excessive amounts it acts as a cathartic. It is reported that foaming may be caused in boiler by more than 50 mg/l of potassium and sodium in water.

CHAPTER - 34
SILICA

Silica is a common constituent of all natural waters present in concentrations ranging from 1 mg/l to 100 mg/l. Silica exists in water, in two forms, namely, reactive and non-reactive (ionic and colloidal respectively). Ionic or reactive silica exists as an SiO_2 complex and poorly ionized and therefore cannot be easily removed by the usual anion exchange resins. Colloidal silica if present will cause serious problems. It can lower the efficiency of filtration systems such as reverse osmosis.

Silica is of special importance in steam generation as it forms troublesome scales of silica and complex silicates in the boiler under certain conditions. Moreover some of it is carried over with the steam and deposits on the blades of steam- turbines resulting in the imbalance of the rotor. Under boiler conditions, it combines with calcium to form calcium silicate. With soluble aluminum (if present) it forms an aluminum silicate scale. It also combines with iron to form hard scales. Sometimes silica alone (with out combining with other substances) forms scales. All these scales are collectively called as 'silica scales'. They are very hard, glassy, adherent and difficult to remove. Even very thin scales of silica results in the failures as their thermal conductivities are very low.

Silica is notorious for its volatility and deposition on turbine blades*. Silica is frequently carried over with the steam, forming scales in superheater tubes and on turbine blades. The deposition of silica scales on turbine blades is not uniform; it may deposit on one blade heavily, in another only lower quantities resulting in an imbalance of the rotor and consequently to breakage too. Besides, such silica deposits on steam turbine blades decrease efficiency and necessitates costly downtime for cleaning. The only remedy to eliminate such problems is the reduction of silica in water so that no deposits are formed either in the boiler or on the turbine blades. Passing water after demineralization – softening through highly basic anion exchange resin would result in the complete removal of silica.

Silica can also form deposits in cooling water systems but it is not of much importance as in boiler system. It becomes significant only when the concentration of silica is about 150 mg/l or if the magnesium content and pH values are high.

---

* Silica levels must be kept below 0.005 mg/l for high pressure boilers.

In process waters, silica has little deleterious effects on the products. In the case of certain cellulose products, a high silica content in the process water is reported to increase the ash content of the finished product. It is also reported that a high silica content may give paper "tinny" characteristics. The recommended limits for silica are 50 mg/l for ground wood pulp, 20 mg/l for soda and sulfate pulp, 50 mg/l for kraft pulp (bleached), 100 mg/l for kraft pulp (unbleached) and 20 mg/l for fine papers. The presence of silica beyond a certain limit is reported to be detrimental for rayon manufacture also. The recommended maximum upper permissible level for silica in water for rayon pulp production is 25mg/l. The recommended threshold value for brewing is 50 mg/l.

Removal of silica is always a tedious task. As stated above, reactive silica is poorly ionized and cannot be removed by the usual anion exchange resins. Highly basic anion exchange resins have to be used for the removal of silica. The reactive silica also has the tendency to polymerize to colloidal silica. It is to be mentioned that colloidal silica will not be retained on the filter bed and passes along with water. However colloidal silica can be removed by ultra filtration; or its concentration may be reduced to very low levels by treatment with a combination of chemicals including lime, soda ash and magnesia.

Though silica in water is undesirable for the above reasons, it is to be mentioned that silicates have been used in water treatment as coagulants for the removal of turbidity and iron. It is also used as corrosion inhibitor. Dosage varying from 4.0 - 8.0 mg/l are applied to form protective coatings on pipes to inhibit corrosion of iron or lead.

CHAPTER - 35

SODIUM

Sodium is present in most natural waters from negligible to appreciable concentrations. The sodium content of water varies over a wide range from below 5 mg/l in fresh water to many thousands of milligram per liter in brines. Sea water contains about 10,700 mg/l of sodium. Sodium present in normal concentration do not interfere in industrial process. Due to the extreme solubility of sodium salts, they do not form scales. However sodium becomes important when salinity or total dissolved solids is a consideration in the use of the water. Sodium concentration is an important factor in boiler waters where the proportion of sodium is high in comparison to other cations present. It is also reported that if the concentration of sodium plus potassium in boiler water exceeds 50 mg/l, foaming trouble may occur. Sodium concentrations in boiler are easily removed by blow down. A report by Larson and King[*] indicates that sodium chloride and other sodium salts in water are strong contributors to the rate of corrosion of iron and steel. (For example, in a water with 400 mg/l of alkalinity (as $CaCO_3$) and pH7, the corrosion rate will be zero at 100 mg/l of NaCl, but when the NaCl concentration is 400 mg/l, the corrosion rate will be about 100 mg per sq.cm per day)

Concentration of 1000 - 1500 mg/l of sodium chloride in the process water causes tastes in canned foods, but does not affect the quality in other ways. For brewing, some sodium chloride in water is desirable and necessary. The recommended range is 75 - 150 mg/l of sodium chloride. More than 500 mg/l is reported to be deleterious.

Determination of sodium is sometimes used to indicate the purity of water; for example, in steam condensates, the concentration of sodium indicates whether there is any carryover from boiler system in to the steam. Likewise the completeness of cation exchange in demineralization effluents can be checked by determining sodium concentration.

Sodium is an important consideration with regard to irrigation waters. In agricultural practice, sodium concentration is expressed in terms of "percent sodium". Percent sodium can be defined as the percentage of sodium concentration of water to the total cationic concentration; ionic concentrations are being expressed in milli equivalent per liter (meq/l). When the percent sodium exceeds 60, the water is considered to be unsuitable for irrigation purpose. (See Chapter – 96 "IRRIGATION WATER" for more details).

---

[*] Larson T.E and King R.M, "Corrosion by Water at low flow velocity" JAWWA, 461 (1954).

CHAPTER - 36
SOLIDS (DISSOLVED)

Natural water normally contains mineral matter, organic matter, clay, silt, algae, diatoms, fungi and bacteria. The mineral matter is found in dissolved condition while others commonly present as suspended impurities.

The dissolved solids (mineral matter) consists mainly of bicarbonates, carbonates, sulfates, chlorides and nitrates of calcium, magnesium, sodium and potassium along with traces of iron and manganese. Silica is another important constituent of dissolved solids. Fluoride is also present in very small concentration. Sometimes humic acid and tannin also appear as constituents of dissolved solids. Phosphate is a common constituent of boiler waters.

The concentration of dissolved solids is the major factor to assess its suitability for various purposes. As regards drinking use, water with a total dissolved solid content below 500 mg/l is preferable. Above this concentration the dissolved solids may render the water non palatable; however it is not harmful to the health of consumers. World Health Organization has set a limit of 1000 mg/l, above which dissolved solids may give rise to consumer complaint – this limit is based on taste considerations and not on any health problem.

Dissolved solids in water are highly undesirable for industrial uses. They affect the color and taste of many finished products. They are the causative agents for scale formation. High concentration of dissolved solids tends to accelerate corrosion. They cause foaming in boilers. The limiting concentrations of dissolved solids for various industrial uses are tabulated below.

TABLE – 36.1
LIMITING CONCENTRATIONS OF DISSOLVED SOLIDS FOR VARIOUS INDUSTRIAL WATERS*

Industry and use of water	Limiting concentrations mg/l
Boiler feed water	50 – 3000 (according to pressure)
Brewing, light beer	500
Brewing, dark beer	1000
Brewing and distilling, general	500 – 1500
Canning and freezing	850
Carbonated beverages	850
Confectionary	50 – 100
Dairy wash waters	850
Food equipment washing	850
Food processing, general	850
Ice manufacture	170 – 1300
Plastics, clear	200
Paper manufacture	
Fine papers	200
Ground wood papers	500
Kraft paper, bleached	300
Kraft paper, unbleached	500
Soda and sulfate pulp	250
High – grade paper products	80
Lower – grade products	150 – 200
Rayon manufacture	100 – 200

Water with extremely low dissolved solids is desirable for atomic power reactor cooling systems.

* Reproduced with the kind permission of the Resources Agency of California, State Water Resources Control Board, California from their publication "WATER QUALITY CRITERIA", 1963, edited by J.E. Mckee and H.W. Wolf.

CHAPTER - 37

SOLIDS (SUSPENDED)

The undissolved matter present in a water is usually referred to as "Suspended Solids"*. It is composed of silt, clays, organic matter, algae, bacteria and fungi. Suspended solids occur in water courses due to natural phenomenon such as erosion. However, significant amount of suspended solids is contributed by the discharge of sewage and industrial effluents to water courses.

Suspended solids are very objectionable in rivers for many reasons. Suspended solids containing much organic matter may cause putrefaction and consequently the stream may be devoid of dissolved oxygen. Mineral and organic suspended matter can lead to silting and if the stream bed is blanketed, destruction of plant and animal life may occur.

For drinking and other domestic uses, presence of suspended matter in water is highly undesirable due to aesthetic considerations and bacteriological safety. Though there has been no limit for suspended solids set by any national agencies for drinking water, it is indirectly controlled by the limits on turbidity (5 units).

Suspended solids interfere with many industrial processes. Especially they cause foaming in boilers and incrustations on equipment. Suspended matter in water is troublesome in cooling waters, as they form clogging deposits. In textile processing, suspended matter is highly objectionable as it is the major cause of staining and streaking. The recommended maximum limit is 5 mg/l. However, even a lower limit is required for some purposes such as package bleaching and package dyeing. Water for steel manufacturing should not contain more than 25 mg/l of suspended solids.

As regards paper industry, suspended solids in water is detrimental in process waters, because it decrease brightness, affect colors, interfere with texture and uniformity, clog wire screens and favors growth of slimes. Suspended solids is also undesirable in waters intended for beverage production, dairy, laundering and power plants.

* Note: One should not confuse between the terms "suspended solids" and "settleable solids". Settleable solids is a fraction of suspended solids that will settle under quiescent conditions, due to the influence of gravity. The coarser suspended solids with a specific gravity greater that of water only will settle.

CHAPTER - 38

SULFATE

Sulfate occurs naturally in water as a result of leachings from gypsum and other common minerals. Sulfate ions are relatively abundant in natural waters ranging from a few to several thousand mg/l. In addition, sulfate is added to water in several treatment processes. The sulfate content of municipal and industrial water supplies is usually increased during clarification, as alum (aluminium sulfate) is added as a coagulant.

High concentration of sulfates is undesirable in potable waters as it imparts a bitter taste to water. Sulfates as magnesium sulfate causes laxative effects particularly in hot climates.

Sulfate in industrial waters causes potential scaling problems. Sulfate is often a constituent of boiler deposits and heat exchanger scales. Sulfate if present as calcium - sulfate is highly detrimental to boilers and heat exchangers. When waters containing calcium sulfate is used in boilers, it does not precipitate until the water is concentrated. It then crystallizes out as a hard and adherent scale on the surface of the boilers which is very difficult to remove even mechanically. All sulfate scales are difficult to remove as they do not respond to acid cleaning. As the thickness of the scale increases, the boiler plates may become seriously overheated and distorted.

Sulfate when present as magnesium sulfate is highly soluble and classified as a non-scale forming constituent. However it may react with chlorides and nitrates under boiler conditions leading to the precipitation of magnesium hydroxide. Though magnesium sulfate is not corrosive, if present in fair quantities in conjunction with chlorides and nitrates of other metals, the risk of corrosion cannot be precluded.

Even sodium sulfate is reported to occur in superheater and turbine blade deposits.

Besides scaling problems, sulfates are also indirectly responsible for corrosion. The micro-organisms namely sulfate reducing bacteria are capable of reducing sulfate to hydrogen sulfide (under anaerobic conditions deriving energy from organic matter obtaining oxygen from sulfate). The produced hydrogen sulfide attacks metals. If sufficient iron is present, under moderate reducing conditions, iron sulfides may also be precipitated.

$$H_2SO_4 + 8H^+ \rightarrow H_2S + 4H_2O$$
$$H_2S + Fe^{2+} + 2OH^- \rightarrow FeS + 2H_2O$$

(Refer Chapter - 17 "Hydrogen Sulfide" for further information)

Despite such demerits, sulfates are desirable in certain industrial processes. The presence of calcium sulfate is advantageous in producing desired flavors in the brewing industry. Infact, waters devoid of sulfate are added with calcium sulfate. (See Chapter – 55 "Brewery")

The State Water Resources Control Board of California has set the following threshold limits for sulfate in industrial water supplies.

TABLE – 38.1
THRESHOLD LIMIT FOR SULFATE IN INDUSTRIAL WATER SUPPLIES

Industrial Process	SO_4 mg/l	$CaSO_4$ mg/l	$MgSO_4$ mg/l	Na_2SO_4 mg/l
Brewing	–	100 - 500	100 - 200	100
Carbonated Beverages	250	–	–	–
Ice Making	–	300	130 - 300	300
Milk Industry	60	–	–	–
Photographic process	–	100	–	–
Sugar making	20	–	–	–
Textiles	100	–	–	–

Sulfates are of considerable importance in the treatment of sewage and waste waters since they are indirectly responsible for odor and sewer corrosion problems. Sulfates are reduced under anaerobic conditions to hydrogen sulfide which causes odor and is oxidized to sulfuric acid and corrode sewer.

CHAPTER - 39

SULFITE

The occurrence of sulfite in natural waters is rare. It is commonly found in boiler waters and cooling waters, as sodium sulfite is added as an oxygen scavenger to protect the systems from corrosion. Sodium sulfite is used extensively to remove the final traces of oxygen that remains in water after physical deaeration. Sodium sulfite reacts with oxygen as shown below* and remove the residual oxygen present.

$$O_2 + 2Na_2SO_3 \rightarrow 2Na_2SO_4$$

A slight excess of sodium sulfite (than the requirement) is always maintained in boiler waters. An excess dosage up to 30 mg/l is frequently employed but higher dosages are detrimental as sulfite can break down to yield sulfur dioxide at higher temperature and boiler pressure. Sometimes, sulfite is converted to hydrogen sulfide in the presence of alkali which is highly corrosive. Hence the sulfite concentration has to be kept as low as possible.

* While this reaction takes place rapidly at boiler temperatures it is much slower at lower temperature and hence sodium sulfite is not used for the removal of dissolved oxygen in cold waters.

CHAPTER - 40

SURFACE ACTIVE AGENTS (SURFACTANTS)

The increased use of synthetic detergents for general cleaning in houses and industries makes the possible presence of surface active agents in many water courses. Surface active agents commonly referred as 'surfactants' are active ingredients of synthetic detergents.

Surface active agents are of three types:

1) *Anionic*
2) *Cationic and*
3) *Non ionic*

Anionic surfactants when dissolved in water ionizes so that the major part of the molecule is the anion. Cationic surfactants ionize so that the major portion of the molecule is the cation. Non ionic surfactants are compounds that do not ionize when dissolved in water.

Among the surfactants used commercially, about 80 percent is only anionic.

The commercial synthetic detergents contain any one of the above surface active agents (20 to 40%) as the basic ingredient. This apart, they contain builders and other fillers such as phosphates (30 to 50%), sodium sulfate (20%), fatty acids, bleaching agents (such as sodium per borate) and sodium carboxy methyl cellulose (1%) to carry out different functions. When synthetic detergents are discharged, not only surface active agents exhibit their effects, but the above mentioned builders and fillers also act in their own characteristic way. For example, phosphates are considered as the principal agents for algal bloom in surface waters.

The presence of surfactants in water is highly undesirable as they are toxic to fish and other aquatic organisms. Water plants are also adversely affected by syndets. As these compounds when present in even very low concentration are able to produce foam which hinders the penetration of oxygen and aeration of river water is prohibited. Similarly purification of sewage by aeration also becomes difficult. Even though foaming does not occurs a film may form over the surface of the water delaying gas exchange with the atmosphere.

Surfactants in drinking water are highly objectionable owing to their tendency to produce unpleasant tastes and odors if present even in extremely small concentration. Besides, corrosion of domestic equipment and plumbing is also attributed to the presence of surface active agents in water.

It is reported that surfactants interfere with the coagulation and flocculation during water treatment. Besides surfactants, phosphates present in detergents also play vital role in retarding coagulation. Chlorination of water containing synthetic detergents may produce unpleasant tastes and odors.

CHAPTER - 41

TURBIDITY

Turbidity is an important parameter for characterizing water quality. It is caused by particulate matter in suspension. It is an expression of optical property of a sample (water or waste water) containing insoluble substances which cause light to be scattered rather than transmitted in straight lines. In most of the waters, turbidity is due to colloidal and extremely fine dispersions. Suspended matter such as clay, silt, finely divided organic and inorganic matter, plankton and other microscopic organisms also contribute to turbidity. Compounds of iron and manganese are important contributors to turbidity.

Turbidity is a characteristic of most surface waters. Ground waters usually are nor turbid. Turbidity is an important consideration in public water supplies because of the following reasons,

1. Aesthetically turbidity is objectionable.
2. Filtration of water is rendered difficult when turbidity increases. Since turbidity shortens the filter runs, satisfactory operation of filter beds becomes impossible.
3. There are chances for the pathogenic organisms to be enclosed in the turbidity causing particles and they may not be exposed to the disinfectant.

Most of the industrial processes need water virtually free from turbidity. Turbidity is highly undesirable in waters for ice making, beverages and brewing, textiles, pulp and paper, steam boilers and turbine operation and other industrial processes.

A summary of the limiting values for turbidity in industrial waters is tabulated below.

TABLE – 41.1

THE RECOMMENDED LIMITS OF TURBIDITY FOR VARIOUS INDUSTRIAL USES OF WATER*

Industry	Turbidity limits
Beverages	1 - 2
Food products	10
Breweries	1 - 10
Boiler feed water	1 - 20 (according to pressure)
Paper and pulp	
Alkaline pulps	25
High - grade paper	5 - 25
Fine writing and book paper	10
Unbleached Kraft paper	100
Bleached Kraft paper	40
Ground wood paper	50
Textiles	
Nitro cellulose	0.5
Rayon	1
Cotton	25
Baking	10
Cooling water	50
Ice making	1 - 5
Tanning	20

In ion-exchange softening, turbidity is of serious concern. Turbidity in water coats the resin particles resulting in poor operation and loss of exchange capacity. Hence special attention is necessary to remove turbidity prior to passing in to ion exchange column. In case turbid waters are passed, special cleaning operations along with regeneration are necessary.

* Reproduced with the kind permission of the Resources Agency of California, State Water Resources Control Board, California from their publication "WATER QUALITY CRITERIA", 1963, edited by J.E. Mckee and H.W. Wolf.

CHAPTER - 42

ZINC

Zinc is commonly found in small quantities in domestic water supplies and industrial waters due to the corrosion of galvanized iron and brass in condensing, cooling and distributing systems. Zinc may also enter into a water supply through the discharge of industrial effluents such as galvanizing wastes, zinc plating wastes, viscose rayon wastes and rubber factory wastes.

Zinc is not reported to exert any deleterious effect on industrial waters. However, a report by Negus* indicates that zinc–bearing water should not be used for preparing acid drinks like lemonade because zinc citrate and other organic zinc compounds that will result may be poisonous.

Zinc has no known adverse physiological effects upon human beings. Infact it is an essential and beneficial element in human nutrition. From aesthetic considerations, however high concentrations (above 5 mg/l) of zinc in domestic water are undesirable. A bitter astringent taste is produced by zinc. Concentrations of zinc above 30 mg/l give water a milky appearance and cause a greasy film on boiling. It is now reported that even 5 mg/l can cause a greasy film on boiling.

* Negus, S.S, "The Physiological Aspects of Mineral Salts In Public Water Supplies" JAWWA, 30, 242, (1938).

CHAPTER - 43

OTHER ELEMENTS

(ARSENIC, CADMIUM AND SELENIUM)

ARSENIC

The occurrence of arsenic in natural waters is rare. It may occur as a result of mineral dissolution, industrial discharges or the application of pesticides. Potable waters usually contain no arsenic or only trace quantities, excepting in certain regions.

Arsenic in drinking water is highly undesirable because of its toxicity. It is a cumulative poison and possesses carcinogenic properties. The WHO International Standards for Drinking Water has prescribed a maximum permissible limit of 0.01 mg/l of arsenic.

CADMIUM

Cadmium salts are usually found in wastes from electroplating industries, textile printing, pigment works, lead mines and chemical industries. These effluents when discharged into water courses, contribute cadmium. It has high toxic potential and even minute quantities of cadmium are responsible for adverse renal arterial changes in human kidneys. Cadmium tends to concentrate in the liver, kidneys, pancreas and thyroid of human beings and animals. The maximum permissible limit of cadmium in drinking water as prescribed by WHO is 0.003 mg/l.

SELENIUM

Seepage through seleniferous soils and industrial effluents contribute selenium to water. The selenium concentration of most waters are generally much less. Concentration exceeding 500 μg/l are rare.

Selenium is a potential carcinogen and causes toxic effects to man and animals. Selenium also causes dental carries to human beings. The maximum limit for selenium in drinking water as prescribed by WHO is 0.01 mg/l.

Arsenic, cadmium and selenium are reported not to interfere with industrial processes. However their presence is highly undesirable in waters intended for food processing and fermentation industries, due to their high toxicity.

CHAPTER - 44

OTHER IMPORTANT PARAMETERS

Hitherto, the impurities present in water have been discussed in detail. In addition, the following parameters have also to be considered before arriving a conclusion on the suitability of water for an industrial process and for designing a suitable treatment plant. These are actually not impurities but parameters that reflect the nature and quality of certain impurities.

1) *Electrical conductivity*
2) *Langelier Saturation Index*
3) *Equivalent Mineral Acidity (EMA)*
4) *Silt Density Index (SDI)*

This apart, the organoleptic properties such as odor and taste are also of prime concern when water is used for food processing and soft drinks manufacture.

ELECTRICAL CONDUCTIVITY

Electrical conductivity is a measure of water's capacity to convey electric current. Electrical conductivity of a water is directly proportional to its dissolved mineral matter content[*]. The unit of conductivity is micro mhos/cm[†] (reciprocal megohm/cm).

Since electrical conductivity determination is very rapid, the quantity of dissolved salts in a water can be ascertained at once. When waters from the same source are periodically examined, the variation in the dissolved solids content is indicated by conductivity measurements. In mixing waters from two or more sources the proportion of each water can be determined using conductivity data. Conductivity is a useful guide to the purity of distilled water and is used for evaluation of steam purity in power industry. (freshly distilled water has a specific conductance of 0.5 - 2 micromhos/cm).

LANGELIER SATURATION INDEX

Langelier saturation index is a measure of the water's ability to dissolve or deposit calcium carbonate and is often used as an indicator to find out whether the water is corrosive

[*] Dissolved gases also contribute to electrical conductivity but silica and organic matter do not contribute.
[†] Also termed as micro Siemens/cm

or scale forming in nature. This index is based on the difference between actual pH value and theoretical pH value which is calculated from total dissolved solids, temperature, calcium hardness and alkalinity.

Langelier Saturation Index, $I_{sat} = pH - pH_s$
$pH_s = (9.3 + A + B) - (C + D)$

where A is the factor for total dissolved solids
B is the factor for temperature °C,
C is the factor for calcium hardness (as $CaCO_3$), and
D is the factor for total alkalinity (as $CaCO_3$).

These factors can be obtained from the following table. If I_{sat} is zero, the water is in chemical balance (i.e) it does not precipitate or dissolve calcium carbonate; if it is a plus quantity, the water is regarded of scale forming in nature. If it is a negative quantity it indicates that the water is corrosive.

TABLE – 44.1
DATA FOR CALCLUATION OF LANGELIER AND RYZNAR INDICES (CALCIUM CARBONATE SATURATION INDEX)

A

TOTAL SOLIDS, mg/l	A
50 to 300	0.1
400 to 1000	0.2

B

TEMPERATURE°C	B
0 to 1.1	2.6
2.2 to 5.5	2.5
6.7 to 8.9	2.4
10.0 to 13.3	2.3
14.4 to 16.7	2.2
17.8 to 21.1	2.1
22.2 to 26.7	2.0
27.8 to 31.1	1.9
32.2 to 36.7	1.8
37.8 to 43.3	1.7
44.4 to 50.0	1.6
51.1 to 55.6	1.5
56.7 to 63.3	1.4
64.4 to 71.1	1.3
72.2 to 81.1	1.2

C

CALCIUM HARDNESS (as $CaCO_3$), mg/l	C
10 to 11	0.6
12 to 13	0.7
14 to 17	0.8
18 to 22	0.9
23 to 27	1.0
28 to 34	1.1
35 to 43	1.2
44 to 55	1.3
56 to 69	1.4
70 to 87	1.5
88 to 110	1.6
111 to 138	1.7
139 to 174	1.8
175 to 220	1.9
230 to 270	2.0
280 to 340	2.1
350 to 430	2.2
440 to 550	2.3
560 to 690	2.4
700 to 870	2.5
800 to 1000	2.6

D

METHYL ORANGE ALKALINITY (as $CaCO_3$), mg/l	D
10 to 11	1.0
12 to 13	1.1
14 to 17	1.2
18 to 22	1.3
23 to 27	1.4
28 to 35	1.5
36 to 44	1.6
45 to 55	1.7
56 to 69	1.8
70 to 88	1.9
89 to 110	2.0
111 to 139	2.1
140 to 176	2.2
177 to 220	2.3
230 to 270	2.4
280 to 350	2.5
360 to 440	2.6
450 to 550	2.7
560 to 690	2.8
700 to 880	2.9
890 to 1000	3.0

Langelier index though not quantitative, is more useful in estimating the water treatment requirements for low pressure boilers and cooling towers.

Similar to Langelier index there are so many indices available, namely, Ryznar Stability Index, Puckorius Practical Scaling Index, Aggressive Index, Stiff-Davis index, Driving-force Index and Momentary Excess. Detailed discussion of these indices are beyond the scope of this book.

EQUIVALENT MINERAL ACIDITY

Equivalent Mineral Acidity (EMA) is an important parameter used to design demineralization plants etc., The sum of the anions which are capable of forming strong acids with H^+ ions is called the equivalent mineral acidity (EMA).

The ions contributing to equivalent mineral acidity are sulfate, chloride and nitrate. If a water containing these ions is passed over a cation exchanger (in the H^+ form), then this latent acidity (EMA) becomes free mineral acidity (FMA).

EMA is usually expressed in terms of calcium carbonate. Hence the sulfate, chloride and nitrate concentrations are first converted into their calcium carbonate equivalents and then summed up.

SILT DENSITY INDEX

Silt Density Index (SDI) is a test used to measure the level of suspended solids in feed water for a reverse osmosis system. This test is also called 'microporous membrane pluggage test' and 'fouling index'. SDI is used as a means of correlating the maximum level of suspended solids allowable in the feed water to fouling in RO systems in order to maintain membrane performance. SDI values less than 4 is recommended for feeding RO systems without treatment (especially for spiral wound elements). This method is based on the rate of pluggage when water is passed through a 0.45 micron membrane filter at a constant applied pressure (30 psi).

CHAPTER - 45

MICRO - ORGANISMS

Micro organisms are found in water in a great variety. A study of micro- organisms is as essential as the study of physical and chemical impurities present in water. They impair the quality of water in their own characteristic way. They may be classified as bacteria, fungi and algae.

BACTERIA

Bacteria are ubiquitous in nature. They are common to surface waters but do not normally present in ground waters. If present in well waters, only in very small numbers. However if the wells are polluted, they too contain a large number of bacteria. They are undesirable both in drinking and industrial waters. Needless to say that some of the bacteria are pathogenic to man. Some bacteria (e.g. Pseudomonas, Flavobacterium and Chromobacterium) can discolor water supplies. Some other bacteria impart tastes and odors. Growth of bacteria in water may interfere with many industrial processes such as the manufacturing of leather products, paper, textiles, breweries and food processing. In food processing / manufacturing industries, bacteriological quality requirements are more stringent than drinking water as bacterial contamination may result in fermentation, change in taste and / or spoilage. Each group of bacteria impair the quality of water in its own way. The prominent bacterial groups with reference to industrial waters are:

Iron bacteria, Manganese bacteria, Sulfate - reducing bacteria, Sulfur – oxidizing bacteria, Slime formers, Proteolytic and Lipolytic bacteria and Gelatin liquefiying bacteria which are discussed below.

IRON BACTERIA

The iron bacteria are of the most important types of nuisance organisms in water. They are generally filamentous and typically found in fresh water. Unlike algae, they are capable of growing in dark and hence they are also seen in pipe lines, reservoirs, stagnant surface waters, wells and springs. They transform soluble compounds of iron to insoluble compounds (ferric hydroxide) which may be deposited in a sheath around the organism (Sphaerotilus) or secreted so as to form stalks or ribbons attached to the cell (Gallionella) or deposit on pipes. The large amount of brown slime so produced will impart a reddish tinge (and an unpleasant odor) to water rendering it unsuitable for drinking and other industrial uses. It is especially troublesome in water intended for boiler feed and cooling. Heavy growths of these organism at times block the passage of water in pipes, valves, nozzles and other parts of water distribution systems, and reduce the flow rate considerably. Their presence may cause pitting and tuberculations in pipes.

In many cooling water systems, iron and manganese bacteria growth causes serious problems necessitating treatment. Some of these growth are matted and fibrous in appearance, some other are dense and resemble inorganic scale. They frequently break loose in large masses and clog nozzles, pumps and recirculating systems etc., as indicated already. This apart, the decaying masses often cause bad tastes and odors in water.

Further, iron bacteria are reported to coat the filter media and ion exchange resins impeding the operation and resulting in a poorly filtered / softened water.

MANGANESE BACTERIA

Manganese bacteria resembles iron bacteria in all respects. Profused growths of manganese bacteria appear black in color. Similar to iron bacteria, their profused growth blocks the passage of water in pipe line and recirculating systems. Also at frequent intervals, they break loose in the form of large masses.

SULFATE-REDUCING BACTERIA

Sulfate reducing bacteria are micro organisms capable of reducing sulfate (or other forms of oxidized sulfate) to hydrogen sulfide. The hydrogen sulfide is a corrosive agent and attack severely the metallic structures and concrete with the formation of 'black water'. These are normally seen at dead ends of mains and the black color is due to ferrous sulfide. Sulfate reducing bacteria are reported to be the most common organism found in the slimes that plug up tube of cooling systems. They form a coating over the surface of the tubes as slime, which acts as an insulator, resulting in reduced heat transfer.

SULFUR OXIDIZING BACTERIA

Waters containing sulfur oxidizing bacteria are frequently seen with thread - like growths. These waters often contain elemental sulfur in finely divided colloidal form. This sulfur may occur due to the oxidation of hydrogen sulfide by bacteria in the presence of dissolved oxygen.

SLIME FORMING ORGANISMS

Many micro - organisms are capable of producing gummy or mucilaginous materials, either as capsular structures or as an extra cellular excretion products. This group includes a wide variety of organisms. Among them the most important ones are iron bacteria, sulfate - reducing bacteria, certain other higher bacteria (myxobacters) and some fungi. The presence of these slime-forming organisms is especially undesirable in water intended for paper mills and food processing industries. These slimes also coat resin particles of ion exchangers and block the surface of the resin resulting in a severe loss of exchange capacity. The bacterial slimes may be destroyed by chlorination of the feed water or by the application of bleach liquors along with regenerant solution. Slimes formed over the resins may also be removed by applying warm caustic soda solution to the resin. After applying caustic soda, the softener should be allowed to stand for several hours to ensure complete removal of slimes and then backwashed to remove the dead slime from the bed.

PROTEOLYTIC AND LIPOLYTIC BACTERIA

Proteolytic bacteria are a group of bacteria that hydrolyze proteins. The prorteolytic bacteria are primarily the aerobic spore - formers (bacillus). Another group of bacteria namely, Lipolytic bacteria hydrolyzes fats and lipids. Hence waters containing these bacteria are undesirable for food processing industries.

THERMOPHILIC BACTERIA

Thermophilic bacteria grow well at the temperatures used in pasteurization, especially when the low - temperature method is followed. These organisms are usually found in milk and are able to survive to pasteurization and can grow at such temperatures. They have an optimum range of 50 to 52°C. The thermophilic bacteria are found in soil and improperly cleaned milking machines and on utensils and pasteurizing equipment. If the utensils and equipments are not thoroughly washed, these organisms collect in the milk films and act as foci for seeding new lots and deteriorate milk.

GELATIN LIQUEFYING BACTERIA

There is a group of bacteria capable of transforming a gel in to liquid and they are collectively termed as "Gelatin Liquefying Bacteria". Their presence is detrimental in waters intended for gelatin manufacture, photographic film manufacture and food processing industries. Enterobacter liquefaciens and Enterobacter cloacae are the important organisms among this group.

FUNGI

Fungi are ubiquitous in nature and are commonly associated with decaying organic matter. The effects of fungus contaminated water for industrial uses are little known. However, fungi are reported to be a factor in the deterioration of wood in cooling towers. Problem of caking of filter sand is also attributed to the fungus growth.

ALGAE

Algae are the most troublesome micro - organisms in water contributing to tastes and odors. The algae are a group of life forms that may be unicellular or multicellular with photosynthetic pigments. Their presence in water is undesirable for a variety of industrial uses – food processing, paper manufacture, dyeing, soft drinks manufacture, laundering and chemical manufacture – a few to name.

Chlorination is the preferred method for killing these organisms. Besides chlorine, copper salts, chromates and zinc salts are also used to control the microbial growth. Organic compounds such as phenol, chlorinated phenol derivatives, naphthenates and a wide variety of patented products are also used. As regards recirculating cooling water system, chlorination is economical and effectively controls biological fouling. Chlorine also destroys the slimes and larvae of certain higher organisms.

PART III

INUDSTRIAL WATERS AND THEIR QUALITY REQUIREMENTS

CHAPTER - 46

BOILER WATER REQUIREMENTS

(See also Chapters 15 & 91, "HARDNESS" and "THERMAL POWER PLANTS")

A chapter on boiler water is included here inorder to provide guidelines for industries. Most of the industries devote much attention to the fire side of the boiler and keep it free from accumulation of deposits which may affect the heat transfer. It is equally important that they should afford utmost care and attention to keep the waterside free from deposits. Such deposits on the water side are potentially dangerous and may result in overheating of the metal and consequently results in tube failures.

Necessary instructions would normally be given by the suppliers of boilers regarding the quality of boiler feed water, boiler water and treatment to be adopted. They must be strictly adhered to. In the following sections, constituents of water, their effects on boiler, boiler water requirements are discussed.

No natural water can be used directly for boilers, as the constituents of water will affect either in one or other way the boiler system and the water needs treatment. The potential problems caused by the water and their effects are listed in Table 47.1.

SCALE FORMATION

The scale is formed mainly by calcium (and magnesium) salts present in water. Sodium salts are highly soluble and do not form scales.

Calcium sulfate and calcium carbonate are the principal scale formers. Calcium-sulfate is the most dangerous as it forms very hard and dense scales and are difficult to remove. The scales formed by calcium carbonate are relatively softer. Magnesium sulfate and magnesium chloride are highly soluble and hence do not form scales. However magnesium hydroxide forms a sludge which is highly undesirable.

TABLE – 46.1
PROBLEMS CAUSED BY WATER

Problem	Effect	Caused by
1. Scale Formation	Deposition over the metallic surfaces and impeding the heat transfer. Responsible for tube failures.	Calcium carbonate, Calcium sulfate, Silica
2. Corrosion	Eating away of boiler metal. Deterioration and equipment failure.	Dissolved oxygen, Carbon dioxide, low pH
3. Priming, Foaming and Carry over	Priming and foaming contaminate the steam. When a boiler is foaming or priming it is difficult to read the true level of the boiler water. Carryover results in the deposition of solids (present in the water) on valves, and engine or turbines.	During operation
4. Formation of Oil Films	Interfere with the free transfer of heat and may result in bulging or even bursting of some of the portions.	Oils and grease
5. Caustic Embrittlement	Metal below the water line becomes brittle and weak. Eventual failure and boiler shut down.	High pH due to highly concentrated caustic boiler water

The other important scale former is silica. While silica is not a troublesome constituent in low pressure boilers, it is the most unwanted substance in water used for high pressure boilers. It forms hard, glassy scales in boilers (and on turbine blades in power generating equipment). The silica scale formation is attributed to the mechanical carry over as well as its volatility (or solubility) in high pressure steam. Silica alone or in combination with aluminium, iron, calcium, magnesium or sodium forms scales, and these scales are collectively termed as 'silica scales' (Refer Chapter - 34 "SILICA" for more details). Suspended matter, turbidity and sediment in boiler water are highly detrimental as they form both scale and sludge. Therefore the turbid waters should be clarified, (filtered if necessary) and then passed on to boilers. Care should be exercised in treating water, as improper treatment may lead to the introduction of calcium- hydroxide which is also a scale former and does not occur in natural waters.

The scale is mainly formed in the feed water piping and boiler tubes. Since scale is deposited inside the piping system, the dia of the pipe is reduced which results in the reduction of flow of water. Due to scale formation, the transfer of heat from the hot gases to water is also reduced. On the contrary, the metal is overheated and may even result in rupture of tubes. The effects of scale become more severe as the boiler pressure increases.

CORROSION

Corrosion is a phenomenon of eating away of boiler metal. Corrosion is the major reason for the deterioration and failure of the equipments leading to expensive shutdowns and / or replacements. The corrosiveness of water is due to its low pH, dissolved gases and some of the salts present in it. Needless to say that low pH induces corrosion. (However, under boiler conditions, concentrated alkalis contributing to high pH are also reported to attack steel). Dissolved oxygen and carbon-dioxide are the principal gases present in water inducing and accelerating corrosion. A solution of oxygen in water is very corrosive to metals such as iron, steel, galvanized iron and brass. Low pH values accelerate the rate of this dissolved oxygen corrosion. (High pH values tend to retard it and that is why the water is made sufficiently alkaline for use in boilers). Carbon dioxide by itself is corrosive in nature and is also an accelerating factor in dissolved oxygen corrosion. Condensate waters are more corrosive due to the presence of oxygen and carbon dioxide. (For more details, refer Chapters – 29 and 7 "OXYGEN (DISSOLVED) and "CARBON DIOXIDE (FREE)". Corrosiveness of water is also attributed to the presence of other gaseous impurities such as ammonia, hydrogen sulfide and sulfur dioxide.

Among the salts present in water, magnesium chloride and sulfate are noted for their corrosiveness. Magnesium chloride especially is corrosive and the action seems to be rather cyclic in nature. The following reactions are reported to take place inside the boiler. Magnesium chloride decomposes into magnesium hydroxide and hydrochloric acid. The latter attacks the boiler metal forming ferrous chloride. This ferrous chloride subsequently reacts with the magnesium hydroxide to form ferrous hydroxide and magnesium chloride. (This type of attack is prevented by the sodium alkalinity in the boiler salines).

Corrosion inside a boiler leads to the pitting or perforation of the metal and loss of mechanical strength due to thinning of metal and / or development of cracks.

PRIMING, FOAMING AND CARRYOVER

Priming, Foaming and Carryover are closely associated terms. Each describes the conditions that cause boiler water to entrain into the steam.

Priming is the more violent discharge of water with steam from the boiler. Priming is similar to that of bumping of water that can be seen when a water is heated rapidly in a vessel. Due to priming, sudden agitation occurs with the result sludge of boiler water is thrown over with the steam. Priming can be reduced considerably by lowering the water level in the boiler and maintaining constant load in boiler.

Foaming is the production of froth and the surface of the water. High percentage of dissolved solids, excess alkalinity and presence of oil are reported to cause foaming. The resulting foam may stand above the water surface or it may fill the entire steam space, paving way for contamination of steam. When a boiler is foaming or priming, it is difficult to read the true level of the boiler water.

Boiler water solids are also carried over in the moisture mixed with steam when there is no priming or foaming. This is known as 'Carryover'. In addition to mechanical factors (such as boiler design, high water level and overloading) chemical factors also contribute to carryover. Among them, the amount of suspended solids and alkalinity of boiler water are very important. The carryover interferes operation of the equipment by forming deposits on valves, engines or turbines. These solids erode turbine blades.

OIL FILMS

Some times oil and grease find their way into boiler feed water. They are very bad conductors of heat and form heat-insulating film on boiler tube or plate. The film interferes with the free transfer of heat to water. With the result, overheating takes place at some portions which may lead to bulging (and sometimes even bursting).

Oil and grease of either animal or vegetable origin decompose to form free organic acids and glycerin under boiler conditions. These free acids attack the boiler metal in the same manner as carbon dioxide. The attack is also cyclic in nature. Therefore it is essential to bestow greatest care to see even a trace of oil or grease is not entering into boiler system. Whenever needed, mineral oils may be used in place of vegetable (or animal) oils. It is however reported that under certain conditions these oils can also promote an acid attack on iron.

Oils are also reported to cause foaming in boilers.

Oil and grease are normally removed during clarification of water using coagulants such as alum and lime. If clarification is not carried out, the water may be passed through multi-media filters to get rid of oils.

CAUSTIC EMBRITTLEMENT
(Refer also, Chapter - 91"THERMAL POWER PLANTS")

Caustic embrittlement also known as, 'Intergranular Cracking' may be defined as the destructive action of highly concentrated caustic boiler water on metal below the water line. Due to caustic embrittlement, irregular cracks appear where the metal is under strain. In the initial stage, caustic embrittlement is observed as very fine hair-like microscopic cracks in boiler plates or rivets. Such cracks are a potential source of danger as further cracks may develop from their extremities. The metal then becomes brittle and weak and may result in eventual failure and boiler shutdown. The phenomenon of caustic embrittlement was common in riveted boilers, but with the advent of welded, stress-relieved boilers, it has become rare.

Caustic embrittlement can be prevented by maintaining low-hydroxide alkalinity in the boiler water. The most effective method of maintaining low-hydroxide alkalinity is adoption of 'Co-ordinated phosphate-pH control'. In this method controlled addition of phosphate blends (disodium and trisodium phosphates) is made so that on hydrolysis, the solutions maintain sufficient alkalinity to inhibit corrosion, yet do not deposit free sodium hydroxide on evaporation. Avoiding leaks at stressed metal areas is also reported to reduce caustic embrittlement. Older methods of prevention of caustic embrittlement include the addition of inhibitors such as sodium nitrate, quebracho extracts and lignin sulfonates. Such method of addition is seldom practised at present.

To overcome these difficulties described above, the water has to be treated suitably. The necessary treatment to be applied depends upon the operating pressures of boilers. While low and medium pressure boilers require water of stringent quality, high pressure boilers require water of utmost purity. The quality requirements as recommended by American Society of Mechanical Engineers and as given in the book, "The Chemical Treatment of Boiler Water" authored by James W.McCoy* are given below. See Table 47–2, 47 – 3 and 47 – 4. These tables should be regarded only as a general guide. Specific instructions would be normally issued by the boiler manufacturers. Wherever trouble arises, boiler water specialists should also be consulted.

* James W.McCoy, "The Chemical Treatment of Boiler Water" – 1981, Chemical Publishing Co., New York.

TABLE - 46 – 2. *

ASME CONSENSUS TABLE 1) Suggested water chemistry limits. Industrial watertube, high duty, primary fuel fired, drum type

Makeup Water Percentage: Up to 100% of feedwater, Conditions: Includes superheater, turbine drives or process restriction on steam purity

Saturated steam purity target: See tabulated values below

	Drum Operating Pressure (psig) / (MPa) (1) (11)							
Feedwater (7)	0-300 (0-2.07)	301-450 (2.08-3.10)	451-600 (3.11-4.14)	601-750 (4.15-5.17)	751-900 (5.18-6.21)	901-1000 (6.22-6.89)	1001-1500 (6.90-10.34)	1501-2000 (10.35-13.79)
Dissolved oxygen ppm (mg/l) O_2 - measured before chemical oxygen scavenger addition (8)	<0.007	<0.007	<0.007	<0.007	<0.007	<0.007	<0.007	<0.007
Total iron, ppm (mg/l) Fe	≤0.1	≤0.05	≤0.03	≤0.025	≤0.02	≤0.02	≤0.01	≤0.01
Total copper, ppm (mg/l) Cu	≤0.05	≤0.025	≤0.02	≤0.02	≤0.015	≤0.01	≤0.01	≤0.01
Total hardness, ppm (mg/l) as $CaCO_3$	≤0.3	≤0.3	≤0.2	≤0.2	≤0.1	≤0.05	ND	ND
pH range @ 25°C (77°F)	8.3-10.0	8.3-10.0	8.3-10.0	8.3-10.0	8.3-10.0	8.8-9.6	8.8-9.6	8.8-9.6
Chemicals for preboiler system protection	NS	NS	NS	NS	NS	VAM	VAM	VAM
Nonvolatile TOC, ppm (mg/l) C (6)	<1	<1	<0.5	<0.5	<0.5	<0.2	<0.2	<0.2
Oily matter, ppm (mg/l)	<1	<1	<0.5	<0.5	<0.5	<0.2	<0.2	<0.2
Boiler Water								
Silica, ppm (mg/l) SiO_2	≤150	≤90	≤40	≤30	≤20	≤8	≤2	≤1
Total alkalinity, ppm (mg/l) $CaCO_3$	<350(3)	<300(3)	<250(3)	<200(3)	<150(3)	<100(3)	NS(4)	NS(4)
Free hydroxide alk. ppm (mg/l) $CaCO_3$ (2)	NS	NS	NS	NS	NS	NS	ND(4)	ND(4)
Specific conductance (12) (μmhos/cm) (μS/cm) @ 25°C without neutralization	5400-1100(5)	4600-900(5)	3800-800(5)	1500-300(5)	1200-200(5)	1000-200(5)	<150	<80
Total Dissolved Solids in Steam (9)								
TDS (maximum) ppm (mg/l)	1.0-0.2	1.0-0.2	1.0-0.2	0.5-0.1	0.5-0.1	0.5-0.1	0.1	0.1

Notes

NS = Not specified

ND = Not detectable

VAM = use only volatile alkaline materials upstream of attemperation water source (10)

* Reproduced with the kind permission from American Society of Mechanical Engineers from their publication, "Consensus on Operating Practices in Modern Industrial Boilers: 1994

1. With local heat fluxes are >1.5 x 10^5 Btu/hr/ft^2 (>473.2 kW/m^2), use values for atleast the next higher pressure range.

2. Minimum hydroxide alkalinity concentrations in boilers below 900 psig (6.21 MPa) must be individually specified by a qualified water treatment consultant with regard to silica solubility and other components of internal treatment.

3. Maximum total alkalinity consistent with acceptable steam purity. If necessary, should override conductance as blow-down control parameter. If makeup is demineralized quality water and boiler operates at less than 1000 psig (6.89 MPa) drum pressure, the boiler water conductance should be that in table for 1001 – 1500 psig (6.9 – 10.34 MPa) range. In this case, the necessary continuous blowdown will usually keep these parameters below the tabulated maximum values. Alkalinity values in excess of 10% of specific conductance values may cause foaming.

4. Not detectable in these cases refers to free sodium or potassium hydroxide alkalinity. Some small variable amount of total alkalinity will be present and measurable with the assumed congruent or co-ordinated phosphate-pH control or volatile treatment employed at these high pressure ranges.

5. Maximum values are often not achievable without exceeding maximum total alkalinity values, especially in boilers below 900 psig (6.21 MPa) with >20% makeup of water whose total alkalinity is >20% of TDS naturally or after pre-treatment by lime-soda, or sodium cycle ion exchange softening. Actual permissible conductance values to achieve any desired steam purity must be established for each case by careful steam purity measurements. Relationship between conductance and steam purity is affected by too many variables to allow its reduction to a simple list of tabulated values.

6. Nonvolatile TOC is that organic carbon not intentionally added as part of the water treatment regimen.

7. Boilers below 900 psig (6.21 MPa) with large furnaces, large steam release space, and internal chelant, polymer, and/or antifoam treatment can sometimes tolerate higher levels of feedwater impurities than those in the table and still achieve adequate deposition control and steam purity. Removal of these impurities by external pretreatment is always a more positive solution. Alternatives must be evaluated as to practicality and economics in each individual case.

8. Values in the table assume existence of a deaerator.

9. Achievable steam purity depends on many variables, including boiler water total alkalinity and specific conductance as well as design of boiler steam drum internals and operating conditions [(Note (5)]. Since boilers in this category require a relatively high degree of steam purity for protection of the superheaters and turbines, more stringent steam purity requirements such as process steam restrictions on individual chemical species or restrictions more stringent than 0.1 ppm (mg/l) TDS turbine steam purity must be addressed specifically.

10. As a general rule, the requirements for attemperation spray water quality are the same as those for steam purity. In some cases boiler feedwater is suitable; however, frequently additional purification is required. In all cases the spray water should be obtained from a source that is free of deposit forming and corrosive chemicals such as sodium hydroxide, sodium sulfite, sodium phosphate, iron, and copper. The suggested limits for spray water

quality are <30 ppb (µg/l) TDS maximum, <10 ppb (µg/l) Na maximum, <20 ppb (µg/l) SiO_2 maximum, and it should be essentially oxygen free.

11. Low pressure boilers frequently use feedwater that is suitable for use in higher pressure boilers. In these cases the boiler water chemistry limits should be based on the pressure range that is most consistent with the feedwater quality.

12. Conversion from ppm (mgl/) TDS values in the ABMA standards [12] used a factor of 0.65.

TABLE – 46.3 RECOMMENDED FEED WATER QUALITY†
(From the book "The Chemical Treatment of Boiler Water" by James W. McCoy)

Pressure (psi)	Silica range (ppm)	Maximum (ppm)			
		Total hardness	Oxygen	Iron*	Copper*
100	15 - 25	75.00	-	-	-
200	10 - 20	20.00	-	-	-
300	7.5 - 15	2.00	-	-	-
500	2.5 - 5.0	2.00	0.030	-	-
600	1.3 - 2.5	0.20	0.030	-	-
750	1.3 - 2.5	0.10	0.030	0.050	0.020
900	0.8 - 1.5	0.05	0.007	0.020	0.015
1000	0.2 - 0.3	0.05	0.007	0.020	0.015
1500	0.3 max	0.00	0.005	0.010	0.010
2000	0.1 max	0.00	0.005	0.010	0.010
2500	0.05 max	0.00	0.003	0.003	0.002
3200+	0.02 max	0.00	0.002	0.002	0.001

* In modern industrial boilers, which have extremely high rates of heat transfer, these concentration should be essentially zero. Similarly, total hardness should not exceed 0.3 ppm $CaCO_3$, even at the lower pressures; suspended solids in the feed water should be zero, if possible.

† Reproduced with the kind permission from M/s. Chemical Publishing Co., Newyork from their publication "The Chemical Treatment of Boiler Water" By James W.McCoy – 1981.

TABLE – 46.4 RECOMMENDED CONCENTRATIONS OF BOILER SALINES†
(From the book "The Chemical Treatment of Boiler Water" by James W. McCoy)

Pressure (psi)	Saturation temperature (F)	Maximum (ppm)						Range (ppm)		
		Dissolved solids	Suspended solids*	Total alkalinity**	Silica	Sludge conditioners		Residual phosphate	Residual sulfite	Residual hydrazine
						Natural	Synthetic			
100	328	5000.00	500	900	250.00	150	15	NR	90-100	NR
200	382	4000.00	350	800	200.00	150	15	40-50	80-90	NR
300	417	3500.00	300	700	175.00	100	15	30-40	60-70	NR
500	467	3000.00	60	600	40.00	70	15	25-30	45-60	NR
600	486	2500.00	50	500	35.00	70	10	20-25	30-45	NR
750	510	2000.00	40	300	30.00	NR	10	15-20	25-30	NR
900	532	1000.00	20	200	20.00	NR	5	10-15	15-20	0.10-0.15
1000	545	500.00	10	50	10.00	NR	3	5-10	NR	0.10-0.15
1500	596	150.00	3	0	3.00	NR	NR	3-6	NR	0.05-0.10
2000	636	50.00	1	0	1.00	NR	NR	1-3	NR	0.05-0.10
2500	668	10.00	0	0	0.50	NR	NR	NR	NR	0.02-0.03
3200+	705	0.02	0	0	0.02	NR	NR	NR	NR	0.01-0.02

* Guidelines for pressures from 100 to 900 psi apply to conventional field-erected boilers with moderate rates of heat transfer, say, 50,000 Btu/h/ft^2. At high rates characteristic of packaged boilers, large amounts of insoluble material cannot be managed effectively by a dispersant presently available.

** Zero alkalinity refers to hydroxide ion, i.e., (2P – M). There is always some alkalinity produced by ammonia, hydrazine, morpholine, or other bases.

NR = Not recommended

† Reproduced with the kind permission from M/s. Chemical Publishing Co., Newyork from their publication "The Chemical Treatment of Boiler Water" By James W.McCoy – 1981.

CHAPTER - 47

COOLING WATER REQUIREMENTS

Large quantities of water are also used for cooling purposes in industries for condensing steam, for cooling internal combustion engines in diesel plants, and for air conditioning. For cooling, either ground or surface water is employed. Among the constituents of natural water that may prove detrimental to its use for cooling purposes are hardness, suspended solids, dissolved gases, acids, oil and other organic compounds and slime forming organisms. The general requirements for cooling waters are that they should not form scales or other clogging deposits which impede the heat transfer as well as narrow the passage of water; and should not be unduly corrosive under particular conditions of use or reuse.

The principal scale forming constituent in cooling water is calcium carbonate, which is formed by the decomposition of calcium bicarbonate as indicated below.

$$Ca(HCO_3)_2 - CaCO_3 + CO_2 + H_2O$$

The above reaction takes place while the cooling water passes through the system. Magnesium bicarbonate also tends to precipitate in the same way. However it is more soluble(about 20 times) than calcium bicarbonate and relatively smaller quantities only are deposited. Calcium sulfate is generally not a scale former in cooling system due to its higher solubility. If once through system is employed, no difficulties would arise with calcium sulfate. If the cooling water is recirculated and if the concentration of calcium sulfate exceeds 1200 mg/l (as $CaCO_3$), hard scales of calcium sulfate would be formed in the cooling system. Such a special mention to calcium sulfate is made, as most of the cooling waters are treated with acid prior to passing into the cooling system to decompose the calcium carbonate which would otherwise deposit. On treatment with acid, calcium - carbonate is converted into calcium sulfate, carbon dioxide and water. Hence it is apparent that the concentration of calcium sulfate in cooling water exceeds than the original concentration. The acid treatment should be optimized such that the concentration of calcium sulfate in cooling water never exceeds 1200 ppm (as $CaCO_3$).

Apart from mineral constituents, suspended matter in water is troublesome in cooling waters, as they form clogging deposits. Hence turbid waters should never be preferred for cooling purposes.

Iron and manganese bacteria are exceedingly troublesome in cooling waters. Their profused growth often clog the pipes, reducing the quantum of water passing through it, and impeding heat transfer as well. Iron and manganese present in water aids the luxuriant growth of these bacteria and masses and clusters of these bacteria are frequently found in cooling water systems. These masses frequently break loose and may completely block passage of water. Sulfur bacteria also act similarly and they often grow profusely in aerators and ponds handling sulfur waters.

Algae are also nuisance organisms hampering the cooling. Algal growths are very common in cooling towers and ponds. Slime forming organisms are noteworthy in this context. They clog the pipes in cooling system. Hence the prime treatment given for cooling waters is the bactericidal / algicidal treatment which removes the troublesome bacteria and algae.

Corrosion is another major problem experienced with cooling waters. The commonest form of corrosion is due to dissolved oxygen corrosion. This corrosion is accelerated by the low pH values. (Hence care should be exercised in acid treatment of these waters to add only judicious amounts of acid thus keeping the pH value under optimum level. The carbon dioxide present in the water naturally and formed during acid treatment have to be expelled, which otherwise would accelerate the dissolved oxygen corrosion). Corrosion not only eats away the metal but also leads to the formation of 'tubercles' rising above each pit in the metal which resists the flow of water in the cooling system, and heat transfer is also greatly affected. Sulfur bearing waters also exert corrosion by forming metallic sulfides.

It is very difficult to offer a conclusive summary on cooling waters, as cooling intended for diversified applications. In general, the water should have appropriate initial temperatures, should not have the tendency to form heat insulating scales or deposits. It should not be unduly corrosive. Also it should not encourage the growth of iron bacteria, manganese bacteria, sulfur bacteria, slime forming organisms and algae. The following table is one of the most definitive lists of quality requirements for cooling waters.

TABLE – 47.1
LIMITING CONCENTRATIONS FOR COOLING WATER

S. No.	Constituents of Water	Limiting Concentration
1.	Turbidity (NTU)	50
2.	Hardness (as $CaCO_3$),mg/l	50
3.	Iron (as Fe), mg/l	0.5
4.	Manganese(as Mn) , mg/l	0.5
5.	Iron and Manganese, mg/l	0.5

The scale forming tendency / corrosive nature of a water can be predicted by certain calculations, based on the values of i) the total dissolved solids content, ii) the temperature to which the water will be raised, iii) calcium hardness, iv) methyl orange alkalinity and (v) pH of the water. Two indices namely, Langelier index and Ryznar index, based on the above calculations are highly useful in determining the nature of water.

CHAPTER - 48
AIR CONDITIONING
(See also, Chapter - 47 "COOLING WATER REQUIREMENTS")

As excellently defined in "The NALCO Water Hand Book", "*an air conditioning system is nothing more than a mechanism designed to move heat from inside the building to the outside*". Air-conditioning has become the fastest growing cooling water application in recent years. The systems use water for direct cooling of air, closed chilled-water systems, humidification and for dehumidification.

The important element that is essentially required for every air conditioning system is refrigerant. The refrigerant absorbs the heat from the water. The water which gets cooled is circulated in a closed system in a room which absorbs the heat from the surrounding space rendering the room very cool. Since the water acquires heat, it is cooled again by the refrigerant, the cycle continues and the room is kept cool. The refrigerant which becomes hotter by acquiring heat from water is compressed and condensed. The condensers in turn is cooled by the water from cooling tower.

The most commonly used refrigerants are Freon (CCl_2F_2) and ammonia. Carbon - dioxide is also used to some extent.

For study purposes, the air –conditioning system may conveniently be divided in to two components, namely 'open recirculation water loop' and 'closed recirculation water loop' also termed 'chilled water system'. The open recirculating water loop consists of compressor, condenser and cooling tower etc. The 'closed recirculating water loop' comprises of chiller, and piping systems to carry chilled water. Open recirculating water after absorption of heat (from the condensers that condense refrigerant) is cooled by passing to cooling water (or spray pond) which is exposed to the atmosphere. In the closed recirculation water system, water is cooled by refrigerant in a closed chilling system and is not open to the atmosphere. In order to obtain maximum efficiency to the desired level, it is imperative that utmost attention should be bestowed upon the quality of open recirculating water. A constant monitoring of the water quality of closed recirculating water system is also necessary.

Though air conditioning is employed in non-industrial facilities such as hospitals, offices, restaurants, laboratories, shopping centres etc., the potential problems encountered with water are similar to those of in industrial plants, and a suitable water treatment program is essential as for industries. The major problems are listed below:

1) *Scale formation*
2) *Corrosion*
3) *Microbiological growth*
4) *Organic fouling*

In general, water used for air conditioning should be free from turbidity, suspended matter, iron and manganese. The iron content shall not exceed 0.5 mg/l either alone or in combination with manganese. The water shall be non-corrosive. It shall not promote the growth of micro-organism and / or slime formation.

Air conditioning is a process entirely dependent on heat exchange. Hence any scale formation and organic growth would result in poor heat transfer with a consequent reduction in efficiency and increasing energy costs.

Corrosion is as dangerous as scales in the system as it causes failure of the system parts resulting in expensive repair and/or replacement.

Given below are some of the methods to protect the open recirculation systems from scale formation and corrosion.

1) Addition of acid to lower the pH so as to keep the water in a non- scaling condition and addition of chromate to prevent corrosion*.

2) The water may be kept in alkaline condition to prevent corrosion; scale formation is inhibited by the addition of organophosphate scale modifying agents.

Open recirculating systems are always susceptible to air-borne contamination. In addition to scrubbing of gases such as oxygen and carbon dioxide, appreciable quantities of air-borne dust and silt are drawn in to the recirculating water as the system is exposed for

* In systems where the concentration results in a pH of 8.5 or above and sufficient alkalinity is present, the addition of corrosion inhibitors may not be necessary. To protect the metallic parts made up of copper from corrosion, organic nitrogen compounds are added.

cooling. They are highly detrimental as they tend to deposit in areas of low velocity (such as in heat exchangers) and impede heat transfer. Besides, such settled sludge causes corrosion and encourages microbial growths. Dispersants† are normally added so that these materials are not deposited but are kept in suspension and finally removed by blow down.

Microbial growths are very common in cooling systems and it is especially dangerous to air-conditioning plants. When congenial conditions available, aerobic slime forming organisms (including iron and sulfur bacteria), anaerobic bacteria, algae, yeast and molds grow luxuriantly in the tubes resulting in significant loss of heat transfer. Severe cases are reported that the growth is so profuse which completely blocked the flow of water through the tubes. Some of the anaerobic bacteria, (especially sulfate reducing bacteria) are capable of growing in areas where there is no oxygen and produce hydrogen sulfide which is highly corrosive and eat away metals at a faster rate resulting in the formation of holes in the piping structure. Special mention is made about algal growth. Algal growth on cooling tower decks and other areas exposed to sunlight are particularly dangerous in that they plug the drain holes on the tower deck. Constant attention and monitoring is necessary to check these growths and if found, suitable chemicals need to be added to correct these problems.

Though open recirculating water systems are susceptible to so many potential dangers discussed above, the closed recirculating water system is less prone to such problems, as only little (or no evaporation) occurs in a properly operating system. However any leak if present may lead to scale formation. The closed recirculating water also need to be treated as a precautionary measure to protect from metal corrosion, and microbial growth (especially anaerobic microbes such as sulfate reducing bacteria) and from foaming. Normally borate / nitrite based formulations are employed which check scale formation, corrosion and foaming, as they contain antiscalants and anti foaming agents.

There is always a lurking danger from microbes. Though they are unable to exit in hot water systems, they can grow in chilled water systems. Sulfate reducing and nitrite reducing bacteria are common organisms to cite as examples, which by their activity corrode the metal – the corrosion sometimes becomes more severe. Periodical monitoring of water quality (especially analysis for such microbes) would be very useful. Whenever necessary suitable biocides may also be added.

† Dispersants are available as individual produces or in combination with the formulations of multipurpose products intended for scale and corrosion inhibition.

CHAPTER - 49

ALUMINIUM INDUSTRY

The water requirements of the aluminium industry have been summarized in a comprehensive report by the U.S. Geological Survey from which the following material has been abstracted.

Aluminium is made in two steps. The first process is the production of alumina from the bauxite ore. The second process is the reduction of alumina to metallic aluminium by electrolytic action.

In alumina plants, water of high quality is required for the hydrate wash, which constitutes only about 5% of the total water demand. For other water uses, quality is not a serious consideration. Any impurities introduced in the hydrate wash tend to accumulate in the caustic soda solution and impair its function. Fortunately, modern alumina plants employ multiple effect evaporators to concentrate the sodium aluminate solution and the vapor from these evaporators can be condensed and recovered for use in the hydrate wash.

In reduction plants, the major need for water is the sprays used for scrubbing of stack gases to minimize air pollution. Water quality is not a consideration in these sprays. Minor quantities of water are used for boiler feed, engine cooling, and other processes for which high quality is necessary. See Chapters – 46 and 47 "BOILER WATER REQUIREMENTS" and "COOLING WATER REQUIREMENTS".

CHAPTER - 50

AUTOMOTIVE INDUSTRY

Automotive industry includes the manufacture of automotive vehicles including cars, trucks, tractors, autos etc., A wide variety of manufacturing operations are carried out in an automotive industry. These operations are categorized under four groups namely,

1. *Casting (including foundry operations and die castings)*
2. *Machining*
3. *Stamping and Fabrication, and*
4. *Assembly*

Some plants carryout either one or two of the above operations, while integrated plants carryout all the processes. All these processes need huge quantities of water.

Following is the brief discussion of different operations and water uses at these operations and quality requirements of water.

CASTING (FOUNDRY OPERATIONS)

In the foundry, automobile parts such as engine blocks, crankshafts are cast. Water is not at all used directly in such operation. However to arrest dust problems, wet scrubbers are employed for which a significant quantity of water is used. The second main use of water in foundry is for cooling cupola. Since this cooling water is recycled, to protect the cupola from any corrosion, pH of the water is adjusted and corrosion inhibitors are usually added. The third important use of water is in the granulation of slag that is tapped from cupola.

MACHINING

In the machining section, rough cast parts are machined to the correct shape and size. In this section coolants are used while machining. These coolants provide lubricity and cooling to the metal as well as to the tool used to shape. Water is mainly used in this section to makeup the coolants. Only soft waters are used to makeup the coolants. If any hard water is used, calcium ions tend to destabilize the emulsions. Hence removal of calcium and magnesium in the makeup water extends the stability of the soluble oils.

Another major use of water in the machining section is for cooling power house diesel generators, compressors, furnace cooling (if any heat treatment such as annealing is employed) and for air-conditioning. The water used for cooling has to meet the cooling water requirements.

Boilers are also employed in this section to generate steam to drive the hammers in forging operation and to heat various cleaners baths. A water with zero hardness is preferred for boiler use. For other requirements, refer Chapter - 47 "BOILER WATER REQUIREMENTS".

In addition to the above uses water is also used in the machinery section at Paint Spray Booths. Spray booths are invariably installed in every automotive industry. Water is not a constituent of any paint or coating formulation, but actually is intended to protect the floor and other areas from the deposition of excess or overspray of paints. The spray booths are designed in such a way that the excess or overspray of paint falls on a film of water or on a basin of water so that it will not stick on to any place. The major quality requirement of water used in spray booths is freedom from hardness otherwise deposits may be formed inside the tubes carrying water.

STAMPING AND FABRICATION

In this section, parts such as doors, floor pans, trunks, hoods and fenders are manufactured. Water is used in this section for washing and cleaning the parts and for cooling in certain areas. Water used for washing should be free from suspended impurities, hardness, iron and manganese. The same is applicable to cooling waters also. Spray booths are also installed in this section and in that case, water free from hardness is required.

The major use of water is in the plating shop. Water is not only used for cleaning the parts prior to plating but also used as a makeup for plating solutions and for rinsing. For cleaning the parts, soft water has to be used to avoid spotting. For the makeup of plating solutions only the demineralized water has to be used. For further details refer Chapter - 65 "ELECTRO PLATING AND METAL FINISHING".

In this section too, boilers are installed for steam generation and the quality of water used has to meet the boiler water requirements.

ASSEMBLY

This is the section where all the parts produced from other sections are received and assembled to a finished car. The major operations in this section are, welding, bonderizing, painting and assembly. Excepting the final operation namely assembly, water is used in all the three operations either directly or indirectly.

In the welding section, sheet metal is welded to form the frame and floor pan to construct the shell of the car. Water is used to cool the welding tips to protect them from melting. Since closed recirculating systems are installed for cooling, the problems associated with such systems such as fouling and plugging of water lines are common. Besides, the corrosion products may also plug the welding tips. Hence suitable chemical treatments after filtration may be employed. Addition of corrosion inhibitors to minimize corrosion and plugging is also recommended.

Bonderizing is a process by which a metallic phosphate layer (such as zinc phosphate) is formed over the metal surface to protect it from corrosion and to provide an amenable surface for the next primer coating. In bonderizing, water is used for cleaning and rinsing operations. After bonderizing, the shell of the car is given a primer coat by electrostatic process. For the makeup of this water-base paint, only demineralized water has to be used. (Softened water or any other low conductivity water should not be used). Water in this process is not only used as a make-up but also used to cool the paint through a heat exchanger. Besides, this section also contains spray booths for which water is required. See the section "Machining" for the quality requirements of water used in paint spray booths.

Water is also used in this section to test for leaks and for final washing of the car. Needless to say that water free from hardness, iron and manganese is essential for these operations.

In addition to the above major operations, water is used for a variety of purposes. The high pressure boilers installed are intended for power generation and for secondary uses such as heating and drying. Enormous water is required for cooling furnaces, engines and condensers. As indicated already, a substantial portion of water is also used for metal pickling, cleaning, electroplating, anodizing, washing and various other processes. This apart, water is used for drinking, flushing and other general plant uses.

Water used in an automotive industry should be of such quality to satisfy all these requirements. The water used for boilers should meet the quality requirements for high pressure boilers. (See Chapter - 47 "BOILER WATER REQUIREMENTS"). Similarly cooling waters should satisfy all the requirements as prescribed in the Chapter - 48 "COOLING WATER REQUIREMENTS".

Besides mineral matter content, water used in all these processes should be free from color, odor, iron, manganese and biological organisms including iron bacteria which may give rise to profused growth in cooling towers especially.

CHAPTER - 51

BAKERY

Manufacture of bread is a large and important industry. In bread making, nearly equal amount of water as the flour amount is used. But only little attention is bestowed towards the quality of water.

Any potable water is usually suitable for baking purpose. The water should be colorless, clear, odorless, tasteless, free from iron and manganese and of approved bacteriological quality. Hard waters are undesirable for bread making as too much hardness retards the fermentation and too little softens the gluten to produce a soggy bread. Medium hard waters strengthen gluten causing it to retain carbon dioxide better and produce a finer-grained structure. A water of medium hardness (largely calcium sulfate) is desirable as some calcium salts are necessary for proper fermentation. In fact in many bakeries calcium sulfate and calcium chloride are added to dough. The presence of calcium chloride in water is desirable as it has a marked effect in raising the dough. (It is reported that addition of 3.5 g of calcium chloride per 1000 g of flour increases the volume of the loaf by 15.5 percent. Addition of calcium chloride also increases the activity of the yeast and shortens the time of fermentation). As regards magnesium, only waters with a low magnesium content is preferred.

The anions such as sulfates, chlorides, nitrates, nitrites and phosphates do not exert any effect unless combined with a metal that influences fermentation. In fact, they are essential for effective fermentation.

The carbonates present in water neutralize the dough and thus interfere with fermentation. The carbonates of the alkali metals (sodium and potassium) are especially harmful.

The presence of sulfide in water is highly undesirable for bread making. It is reported that a very small quantity of potassium sulfide renders the dough so sticky that it is impossible to make bread from it.

For cake and cracker baking, zero-hardness water is preferable as it is reported that it yields a better and more uniform product.

Zero hardness water is best for cleaning purposes throughout the bakery.

A few threshold values are summarized below:

TABLE – 51.1
THRESHOLD VALUES FOR CERTAIN CONSTITUENTS OF WATER USED FOR BAKING

Constituent	Threshold Values in mg/l
Turbidity	10
Color	10
Taste and Odor	None – low
Iron	0.2
Manganese	0.2
Iron and manganese	0.2
Hydrogen sulfide	0.2

CHAPTER - 52

BEVERAGE PRODUCTION

The water used in beverage manufacture may be classified as:

1) *Water used in making beverage*
2) *Water used for bottle washing and*
3) *Water used for boiler feed water make-up and other general purposes.*

In general, the quality requirements of water used in making beverages are more stringent than those for drinking water. Very low concentration or complete absence of micro-organisms, organic matter, turbidity, color, taste and odor, iron, and manganese are desired. Fruit–juice beverages are particularly susceptible to micro-organism growth. Because of the acid nature of the product, uniformly low alkalinity below 50 mg/l is an important requirement. The flavor of carbonated beverages is adversely affected by high alkalinity. The various threshold values that have been recommended are summarized in Table 52.1.

The quality of water used for bottle washing in a beverage production unit is about 10 to 15 times as that of the water required to produce beverage. The water intended for bottle washing should be free from color, turbidity, odor, iron and manganese. It should be soft, preferably of "zero hardness". The water used for boiler feed water make-up and general purposes should also be soft, preferably of "zero hardness".

TABLE – 52.1
THRESHOLD VALUES FOR CERTAIN CONSTITUENTS OF WATER USED FOR PRODUCTION OF BEVERAGES*

Constituent	Threshold values, in mg/l
Turbidity	1.0 - 2.0
Color	5.0 - 10.0
Taste and Odor	None - low
Alkalinity (as $CaCO_3$)	50
Total dissolved solids	850
Hardness (as $CaCO_3$)	200 - 250
Iron	0.1 - 0.2
Manganese	0.2
Iron and manganese	0.1 - 0.5
Chloride	250
Sulfate	250
Fluoride	0.2 - 1.0
Hydrogen sulfide	0 - 0.2
Organic matter	Infinitesimal
Algae and protozoa	None
Oxygen consumed	1.5

* Reproduced with the kind permission of the Resources Agency of California, State Water Resources Control Board, California from their publication "WATER QUALITY CRITERIA", 1963, edited by J.E. Mckee and H.W. Wolf.

CHAPTER - 53

BOTTLE / UTENSIL / FOOD EQUIPMENT WASHING

Food processing and canning, soft drinks manufacture, alcoholic beverages manufacture and dairies are some of the industries where bottle / food equipment washing is the major operation carried out and large quantities of water are used. (Approximately 10 to 15 times that of process water). The quality requirements for bottle washing, food equipment washing are so stringent and it should match that of process water or else the product may get spoiled. Water intended for bottle / container / food equipment washing should be free from color, turbidity, odor, taste, organic matter, iron and manganese. Water of zero hardness is desirable for washing to prevent 'cloudy appearance' and formation of spots in the bottle. Besides eliminating deposits, zero hardness waters save detergents to a marked extent. Waters containing chlorides not more than 250 mg/l and fluorides 1.0 mg/l are preferable for washing. Iron and manganese reduce the brightness of the bottle and bottles would be dull in appearance after washing.

The bacteriological quality of the water is utmost important. Not only it should meet the requirements of drinking water but it should also be free from saprophytic organisms that may attack particles of food lodged in hard–to–clean crevices or cracks in the equipment.

CHAPTER - 54

BREWERY

History reveals that from the earliest days of brewing, breweries were built near suitable sources of water, and the water in certain areas determined the quality of the beer made. This still holds good, and the quality of water is an important criterion to obtain best quality beer.

Before proceeding to the quality requirements of water for brewing, a brief introduction of the manufacture of beer is given below so that the processes and quality requirements can be correlated well.

In a brewery, beer is manufactured by fermentation of various grains and fruits. The variety of grains and fruits used is wide, changing from country to country. The commonly employed cereal is barley which is malted to develop necessary enzymes and to provide the beer the desired flavor. Other grains such as flaked rice, oats and corn are also used along with malted barley and they are commonly termed as malt adjuncts.

The first step in the production of beer is malt preparation. The barley intended for malting is steeped in cold water[*] and spread out on floors or in special compartments, which is turned over regularly for a period ranging from 5 – 8 days. When germination proceeds, the layer becomes thin. When necessary enzymes are formed, the growth is arrested by heat. During the growth period, an enzyme diastase is formed which is the principal agent in the conversion of dissolved starch[†] to disaccharide maltose. The disaccharide maltose is again broken down to monosaccharide glucose by the enzyme maltase. Yeast readily acts upon glucose to yield alcohol and carbon dioxide.

After malt preparation the production of beer is carried out in three stages, namely, 1) Mashing and Brewing, 2) Fermentation, and 3) Storage, Finishing and Bottling.

---

[*] For steeping grains for malt preparation water quality is specified. Refer Chapter - 72 "MALT PREPARATION" (Page 157)

[†] Starch is a polysaccharide and cannot be fermented. Hence it is disintegrated to monosaccharides which are readily fermented by yeast.

Mashing is the process of macerating the malt with hot water for the extraction of valuable constituents of malt, malt adjuncts and sugars. The malt adjuncts which constitutes about 30 – 40% of the required amount of starch is ground well and taken in a pressure cooker provided with an agitator. The remaining amount is compensated by adding the malt. Sufficient amount of water is added and the mixture is heated to about 110°C (5 lb / in^2) for about 60 minutes. In the pressure cooker the insoluble starch is converted into liquefied starch and the soluble malt starch into dextrin and malt sugars.

In the mean time the remaining major portion of the malt is placed in a mash tub (similar to pressure cooker) and thoroughly mixed with water at a temperature between 38°C – 45°C and allowed to rest. During this rest period, the proteolytic enzymes degrade the malt and other proteins into soluble forms.

After this rest period, the boiling cooker mash is mixed with the malt in the mash tub. Now the temperature is about 65°C – 70°C. The temperature is then raised to about 75°C.

The mixture is then passed onto Lauter tub the bottom of which has slots. The clear liquor termed as "wort" passes through these slots into a jacketed copper kettle. Thus the wort is separated form the insoluble spent grains. For complete recovery of all substances in solution, a spray of decarbonated water at about 75°C is rained through the spent grains. This is called "sparging".

Then the wort in the copper kettle is boiled for about 3 hours. During the period of boiling, hops are added at fixed intervals. The process of boiling along with hops destroys the enzymes, coagulate undesirable proteins, and above all, the wort acquires the desired bitter aroma from hops. After boiling, the wort is strained to remove all solids, especially the spent hops. The clear liquor is then allowed to settle for a while in the hot wort tank and cooled* and passed onto fermentors made up of wood or aluminium or glass - lined steel.

The liquor in the fermentor is then added with yeast and allowed for fermentation. The fermentation takes about 7 – 10 days, after which the beer is run through a cooler into glass–lined tanks, leaving behind the yeast sludge in the fermentor. The beer is cooled to 0°C and stored in the cellar for 3–6 weeks at this temperature. During this period, clarification, separation and precipitation of hard resin occur. Besides, the palatability, of the beer also improves to a significant extent.

* The cooling step not only reduces the temperature, but also allow the wort to absorb enough air to facilitate the start of fermentation.

At the end of the period, the beer is filtered, carbonated and bottled. After bottling, the beer is pasteurized at 70°C (Some beer is not pasteurized, but passed through membrane filter which removes residual yeast cells and harmful bacteria). Then it is moved to a chilling compartment and then sent for packaging.

Although the quality of beer depends largely upon the quality of other ingredients used in its manufacture, quality of water is also very important. The ions present in the water have a marked influence on the brewing, the fermentation, and the quality of the finished product. Water used in brewing, therefore must be carefully analyzed and treated to the desired mineral content. The proper balance between the various mineral salts is important to ensure the desired action of the enzymes in the mashing process, a good fermentation and satisfactorily maturing beer.

In the brewing industry, the use of water is divided into two major categories, namely, process water and production water. Each use has different quality requirements.

Process water is utilized for steam production for heating, or for pasteurization, for cooling, for cleanup operations and for bottle washing. In general the process water should be soft, of low mineral matter content and bacteriologically safe (See Chapter - 47 "BOILER WATER REQUIREMENTS", the Chapter – 47 "COOLING WATER REQUIREMENTS" and the Chapter – 54 "BOTTLE / UTENSIL / FOOD EQUIPMENT WASHING").

The water of production is used in brewing and much of it remains as the major ingredient of beer or ale. Hence its quality is of utmost importance. In general, water should be clear, colorless, odorless, and free from organic matter.

It is essential to understand the effects of salts present in water on the different processes of brewery. Certain mineral salts are harmful. Some of them, however are beneficial in brewing and improve the quality of beer.

HARDNESS

Hardness of water is an important criterion in brewing operations. Soft waters extract more mineral water and organic substances from the grain during steeping, whereas hard waters extract more albuminoid matter. With hard waters steeping takes a longer time. Hence a moderately hard water is preferred. For light ales, a water of medium hardness (largely calcium sulfate) is required. For dark beers, soft waters are preferred. For making light colored beer, presence of calcium sulfate is essential. (Read the discussion under the section "Sulfate of Calcium and Magnesium" in this Chapter).

Besides the above advantages, a sulfate water is useful, as with such waters, cell formation of the yeast is much more vigorous than with soft water. Yeast is well fed when a sulfate water is used. Beers made with sulfate waters taste much and keep better. The proportion of hop resin dissolved is less and the color of the beer is paler.

However waters containing magnesium sulfate are undesirable, as magnesium sulfate is reported to retard the extraction of the malt and impart an unpleasant acrid taste to the product due to formation of excess amyloin.

CARBONATES OF CALCIUM AND MAGNESIUM

It is better for brewing purposes that the water should have a certain amount of hardness. In malting, if the water is too soft, it dissolves too much of the useful substances owing to its solvent action and if it is too hard, the operation is retarded. Water with a moderate concentration of calcium carbonate have a less solvent action than soft water. They regulate the fermentation, facilitate clarification and give good keeping qualities to the beer. Water containing calcium carbonate in excess give the wort an intense red color[*], acting on the malt and especially on the hops and preventing a regular fermentation[†]. Magnesium carbonate reduces the activity of diastase, diminishes the proportion of maltose and of extract and on the other hand increases the proportion of nitrogenous bodies and the hop resin. The beer in this case takes a dark color and an acrid taste. It is very difficult to keep the beer when it has been brewed with water too rich in carbonates. As a matter of fact, beer owes its keeping properties to its acidity, which in this case is neutralized by the carbonates.

SULFATES OF CALCIUM AND MAGNESIUM

Compared to soft waters, waters containing calcium sulfate are beneficial. The sulfate waters have a great advantage over those containing carbonate of calcium in solution, because the sulfate does not precipitate so easily as the carbonate.

[*] Worts prepared with water containing sodium carbonate always have a reddish tint.

[†] Carbonates / bicarbonates precipitate phosphates – an essential food for yeast and thus retards fermentation.

When the water contains a small proportion of calcium sulfate, it neutralizes the acidity of the wort and of the beer. It facilitates the production of peptones and amides during brewing. Further it causes albuminoids to coagulate in readily separable form during the boiling of the wort, resulting in a clear and good-keeping product. Hence if the water contains low sulfate hardness, calcium sulfate is added in quantities ranging from 200 to 250 mg/l.

IRON SALTS

Iron is detrimental in production water. When waters containing a small quantity of iron, it may be used in brewing, but it is very preferable that it should contain no iron at all, as the oxide of iron forms an insoluble compound with the diastase which interferer with the manufacture of beer.

In malting, the barley is colored brown by iron salts.

Beer brewed with a water containing iron more than 0.1 mg/l, usually acquires a greenish color which is the result of formation of a black tannate of iron. Moreover, ferruginous waters, may give the beer an unpleasant taste and odor.

CHLORIDES

The chloride ion renders the beer more mellow and sweet, but it may be harmful in concentrations of over 100 mg/l or perhaps over 60 mg/l. Water containing sodium chloride is undesirable as it interferes with the germination of grains. Some authors state that calcium chloride in water diminishes the yield of the yeast , while few others indicate that water containing calcium chloride increase the malto- dextrins or combinations of maltose , and may be used in breweries to produce beers rich in dextrin.

AMMONIA, NITRITES AND NITRATES

Ammoniacal compounds are formed by the decomposition of organic matter in water. The water supply of a brewery must be completely free from ammoniacal salts.

Nitrites and nitrates result from the oxidation of organic matter. These salts favor the development of Renard ferment which is a disease of the beer, spoiling its keeping qualities.

Nitrates have an effect on the yeast, interfering with the process of fermentation. Nitrates are partially reduced to nitrites during fermention, which in turn are poisonous to yeast. The wort in the presence of nitrates derives an intense reddish color and the beer acquires an acrid taste. Nitrogen compounds in general lead to cloudy appearance of bottled beer. They also give a bad taste to beer when present in large quantities.

The presence of nitrites is highly undesirable because they cause an invasion of the malt by molds and bacteria. This apart, as stated above, nitrites are poisonous to yeast. The preservation of beer is also reported to be poor if nitrites are present in the production water.

The nitrites are particularly harmful when the water contains along with nitrate, ammoniacal salts, nitrites or organic matter as the combined action is reported to produce unwanted reactions affecting the quality of the final product.

ORGANIC MATTER

The presence of organic matter in water is highly undesirable. When the grain is steeped in water containing organic matter, the tendency to mouldiness is favored with the resultant defective malting. It also poisons the yeast and suppresses its activity in the brewing operation. It is reported that when wort is boiled with water containing organic matter, beer acquires a much deeper color, its taste is changed and keeping qualities spoiled. Even in washing the casks and brewery utensils, it is necessary to use water free from organic matter, otherwise the organic matter and bacteria may adhere to the casks and utensils which, may deteriorate the quality of beer.

MICRO – ORGANISMS

Not only mineral and organic matter present in water but micro-organisms are also harmful in brewing. In their presence, the beer acquires a disagreeable taste and its keeping qualities are destroyed. Lactic acid produced by the lactose fermenters usually diminishes the activities of the yeast. That is why the stringent limit of zero for the MPN of coliform is prescribed by water quality authorities for water intended for brewing, as coliform organisms are the best known lactose fermenters and commonly found in water. The process of brewing entirely depends on yeast activity and normally organisms harmful to yeast are contributed by water. It is indispensable therefore, to have water free not only from organic matter but from mico-organisms too.

As regards the viable colony count, hundred colonies per ml is the maximum limit for water used in fermentation industry.

Sometimes water used in brewery is chlorinated to ensure complete sterilization. In such cases, the water has to be dechlorinated by passing through activated carbon filter or else the residual chlorine would destroy some of the organic materials used for flavoring. The presence of certain microscopical organisms is also undesirable as they deteriorate the quality of beer. For example, Beggiatoa alba in water usually gives off characteristic odor of hydrogen sulfide.

It is reiterated again that absolutely clean waters only have to be used in brewery. The vats, casks and all the brewery utensils have to be washed thoroughly.

The various threshold limits for the constituents of water used for brewing are tabulated. (Table - 54.1)

If the beer produced is bottled the cleanliness of the bottles is very important. Prior to filling they have to be cleaned scrupulously with good quality water. For quality requirements of water intended for bottle washing, see Chapter - 53 "BOTTLE / UTENSIL / FOOD EQUIPMENT WASHING".

TABLE – 54.1

RANGE OF LIMITING VALUES FOR THE CONSTITUENTS OF WATER USED IN BREWING*

Constituent	Range of Recommended Threshold or limiting values in mg/l
Color	0 – 10
Turbidity	0 – 10
Taste and Odor	None-low
Total dissolved solids	500 – 1500 300 (for anyone substance)
pH value	6.5 – 7.0
Alkalinity as $CaCO_3$	75 – 80 (light beer) 80 – 150 (dark beer)
Iron	0.1
Manganese	0.1
Iron and Manganese	0.1
Carbonate	50 – 68
Nitrite	0
Nitrate	10
Chloride	60 – 100
Silica	50
Hydrogen sulfide	0.2
Fluoride	1.0
Calcium	100 – 200 (light beer) 200 – 500 (dark beer)
Calcium sulfate	100 – 500
Calcium chloride	100 – 200
Magnesium	30
Magnesium sulfate	50 – 200
Magnesium chloride	50 – 200
Sodium chloride	275 – 500
Sodium sulfate	100
Sodium carbonate	100

* Reproduced with the kind permission from The Resources Agency of California, State Water Resources Control Board, from their publication "WATER QUALITY CRITERIA" 2nd Edn, 1963

CHAPTER - 55

CARBON BLACK INDUSTRY

The term carbon black describes a rather wide spectrum of finely divided carbonaceous pigments which are derived from the pyrolysis of oils and hydrocarbon gases. Carbon black is chiefly used as a reinforcing agent and compounding agent for rubber, particularly for motor vehicle tires and aircraft tires. It also finds wide use in plastics, paints, inks and protective coatings. Addition of about 1 to 2% carbon black in these products is reported to minimize significantly the adverse effects of UV radiation (sunlight) on such materials.

The principal types of carbon black are lampblack, channel or impingement carbon, thermal black, acetylene black and furnace carbon. Furnace carbon is the most widely employed material.

Two methods are employed for the manufacture of carbon black, namely, contact method and furnace process. In the contact method, water finds use only in the final stage (i.e.,) pelleting of finished carbon black. Water is added to form a slurry prior to pelleting. As this water is incorporated in the finished product, the water should contain very little dissolved mineral matter only.

In the furnace process, carbon black is produced by the incomplete combustion of hydrocarbons at 2200 to 3000°F. in this process, practically all the water is used only for quenching. As most of the quenching water is incorporated into the product, the dissolved solids content of the water used for quenching must be very low.

CHAPTER - 56

CERAMIC INDUSTRY

In the manufacture of ceramics, water is used for processing, boiler feed and general purposes. The process water should be colorless and clear. It should be free from iron and manganese. It should also be free from organic growths such as algae, crenothrix (iron bacteria) and sulfate reducing bacteria. Water with low mineral matter content is desirable for ceramic manufacture. Water of high mineral content are usually troublesome. Calcium and magnesium and sulfate ions have a profound flocculating effect on the colloidal portion of the clay which affects the viscosity and workability of the "slip"*. Flocculating materials not only found in water but also in the clay materials itself used for ceramic manufacture. Calcium, magnesium and sulfates are commonly present in water. Calcium sulfate in the form of gypsum may also be present in solid ingredients. Hence it is desirable to keep the levels of these ions in water in minimum concentration.

The commonly employed deflocculating agents are mixtures of sodium silicate and sodium carbonate (in various proportions). They react with calcium and magnesium and precipitate them as relatively insoluble carbonates and/or silicates. Barium carbonate is used to precipitate sulfate as barium sulfate. During the addition of barium carbonate, care should be exercised to add only the required quantity, as barium has a greater flocculating action than calcium and magnesium.

* Slip – mixture of water with clay

CHAPTER - 57

CHEMICAL INDUSTRIES

Chemical industry is the major user of water. More than 10,000 products are stated to be manufactured by the chemical industry. In the manufacture of chemicals, large quantities of water are used for a great variety of purposes. In addition, considerable quantities of water are also used for boiler feed and cooling.

PROCESS WATERS

Water itself is used as a raw material and a portion becomes a part of the finished product. Hydraulic conveying, hydraulic classification, washing of products, purification by crystallization equipment cleanup and gas scrubbing are some of the major uses. The requirements for process water vary widely, but in general these waters should be colorless and clear, free from iron*, manganese, hydrogen sulfide and organic growths. The requirement of any additional treatment depends on the chemicals being manufactured, the tolerances for various impurities and the composition of the process water. In the production of certain chemicals, calcium and magnesium salts are objectionable, while sodium salts may not be objectionable. For the production of such chemicals, softened water is used. For the manufacture of pure chemicals, drugs and pharmaceuticals, demineralized water is essentially required. In the refining of salt, and manufacture of chemicals from brines and sea water, elaborate treatment (starting from coagulation followed by filtration, iron removal, calcium and magnesium removal) is required to obtain a pure water.

BOILER FEED WATERS

Most of the chemical manufacturing processes and operations require large quantities of steam for heating, processing and generation of electricity. In the past, all these steam requirements were accomplished by low-pressure boilers, for which the water quality requirements were not stringent. The modern tendency has shifted towards high-pressure boilers. But waters intended for high-pressure boilers have to meet stringent specifications.

* Iron is the most undesirable constituent. It is reported to catalyze chemical reaction, discolor the finished product, or precipitate to produce a turbid product.

(Refer Chapter – 47 "BOILER WATER REQUIREMENTS"). Therefore treatment of the boiler water includes practically all the unit operations of water treatment, such as coagulation, settling, filtration, aeration, iron and manganese removal, demineralization and silica removal. Treatment for the removal of carbon- dioxide and dissolved oxygen is also essential.

COOLING WATERS

The greatest use of water in chemical industry is for cooling the process equipments. Many chemical reactions are exothermic and generate heat. The reaction vessel is cooled so that the temperature is controlled to the desired limit. The quantity of cooling water varies from industry to industry. While some of the industries use a fraction of the total water consumption for cooling, most of the industries require as large as 90%. Quality requirements of water and treatment therefore vary from industry to industry. In the cooling water is used once and then run to waste, no treatment is normally necessary. Or simple treatment such as intermittent chlorination or acid treatment may be required. If the cooling water is reused by recirculation through cooling towers or spray ponds, they have to be treated to protect from scale formation and organic growths. (For further details refer Chapter - 48 "COOLING WATER REQUIREMENTS").

CHAPTER - 58

CIDER MAKING

(See also, Chapter - 53 "BEVERAGE PRODUCTION")

Cider is the fermented drink prepared from apple juice. All the quality requirements of water specified for the beverage production are applicable to cider making also. Since most of the apple growers themselves are cider makers, they do not show any serious concern about the quality of water. But water quality plays a vital role in the quality of cider, its taste and its keeping quality. Water quality is the prime reason why some ciders are better than others. The nature and proportion of the salts present in water determine the quality of cider.

In cider making the apples are squeezed in cider-presses to obtain juice. However the pulp still contains a large quantity of sugary juice which is extracted with water. Water used in cider making should be colorless and clear, odorless and tasteless. It should be of approved bacteriological quality.

Hardness in water is detrimental for cider making as it combines with the apple juice and produce different types of compound that deteriorate the quality of the cider. Similarly the presence of iron, manganese and ammonia are objectionable. When water containing iron is used for making cider, it imparts a black tint. Ammonia is reported to impart brownish tint. This discoloration is technically termed as "draw". In view of these disadvantages, water used for making cider should also be free from iron, manganese and ammonia.

CHAPTER - 59

CONFECTIONARY

Water used in making confections should have a quality suitable for drinking and should be free from constituents that interfere with the complex chemical and physical reaction of the processes. Soft waters are reported to be better for making hard candies. (water quality similar to that of distilled water, with a pH value below 7.0, may favor the inversion of sucrose sugar to produce a sticky candy).

Summarized below are the recommended threshold concentrations of various constituents of water to be used for making confections.

TABLE – 59.1
THRESHOLD CONCENTRATIONS OF CONSTITUENTS OF WATER USED IN CONFECTIONARY

Constituent	Range of Recommended Threshold Values in mg/l
Taste and odor	Low
Iron	0.2
Manganese	0.2
Iron and Manganese	0.2
Total solids	50 - 100
Hydrogen sulfide	0.2
pH Value	> 7.0

CHAPTER - 60

COPPER INDUSTRY

Copper is the metal that requires more water than any other metal for the concentration of ore and reduction of the concentrate to metal. It is reported that about 70,000 gallons of water is used for mining and concentrating a tonne of copper ore, while about 30,000 gallons is stated to be required for the reduction to metallic copper and refinement.

At the mining area, water is used in open-pit mines to control silica dust. Quality is not an important criterion for this purpose. However, the water must be free from any petroleum products as it interfere with the subsequent concentration of the ore by flotation method.

There are two types of copper ores, namely, sulfides and oxides. Sulfide ore is concentrated by floatation technic. Any water (even highly mineralized) can be used for this purpose, provided it should be free from acid and mud. The essential requirement is that it should be free from petroleum products which may adhere to ore and change its specific gravity making it difficult to separate by floatation process. Since oxide ore cannot be separated by floatation technic it is treated with water containing sulfuric acid and leached out. For such leaching, a water of lower mineral content, low pH value and low alkalinity is desirable. The leachate containing copper sulfate in solution is then precipitated by passing through beds of sponge iron or iron scrap.

Copper smelters and refineries require water of good quality. At electrolytic refineries, impure copper is made as anode, while pure copper sheet as cathode. The anode containing about 98% copper is deposited over pure copper cathode during electrolysis. The electrolyte is copper sulfate solution containing sulfuric acid. To prepare this electrolyte bath, high quality water is essential.

The copper cathodes are then rinsed with water. For rinsing, only high purity water is required. Distilled water or demineralized water alone has to be used.

Copper smelters and electrolytic copper refineries use considerable water for boiler feed and cooling purpose. For their quality requirements, refer Chapter - 47 "BOILER WATER REQUIREMENTS" and Chapter – 48 "COOLING WATER REQUIREMENTS".

CHAPTER - 61

DAIRY

Dairy is an industry where milk is processed and various milk products are manufactured. At dairies, raw milk is received from various stations and processed. The processing involves pasteurization, standardization, bottling etc., Raw milk is usually received in cans and tankers and they are emptied and cleaned immediately in the place itself. The milk is then subjected to pasteurization, cream separation and then bottled.

Water is used in the dairy primarily for the cleaning of pasteurizing equipment, for rinsing and cleaning tank trucks and cans, and for bottle washing. The water employed in dairies should be colorless and clear. It should be free from any taste and odor and also free from chemicals that might leave harmful residues. The water should also be free from copper, and very low in iron and manganese and other metals in as much as these metals function as catalysts for the oxidation of fatty matter and for the hydrolysis of milk constituents. Hence they may cause fatty, oily, or fishy tastes in the milk products. Water of zero hardness is preferable in dairy. Hard water forms scales in boilers, water heaters, piping and fixture. It is also troublesome in the pasteurizers, as the temperature of the water used is around 80°C. Hard waters are troublesome in can washing too as they lead to the formation of "milk stone". The recommended limits for chemical and physical constituents are tabulated as follows:

TABLE – 61.1

RECOMMENDED LIMITS FOR CONSTITUENTS OF WATER USED IN DAIRY*

Constituents	Range of Recommended Threshold Values, in mg/l
Color	None
Odor	None
Dry residue	Less than 500 mg/l
Organic matter, in terms of $KMnO_4$	Less than 12 mg/l
Ammonia, as NH_3	Traces only
Nitrite	None
Nitrate as N_2O_5	Less than 30 mg/l
Chloride	Less than 30 mg/l
Sulfate	Less than 60 mg/l
Hardness, as $CaCO_3$	Less than 180 mg/l
Iron	0.1 to 0.3 mg/l
Manganese	0.03 to 0.1 mg/l

* Reproduced with the kind permission of the Resources Agency of California, State Water Resources Control Board, California from their publication "WATER QUALITY CRITERIA", 1963, edited by J.E. Mckee and H.W. Wolf.

A sterile water is desirable in dairy. In the absence of sterile water, the water has to be disinfected suitably to meet the following requirements.

TABLE – 61.2
BACTERIOLOGICAL REQUIREMENTS OF WATER USED IN DAIRY

Parameter	Limit
Standard Plate Count / ml on agar at 37°C for 24 hrs	Not to exceed 50
Coliform bacteria	- 0 -
Proteolytic bacterial count / ml	Not to exceed 5
Lipolytic bacterial count / ml	Not to exceed 5
Proteolytic and Lipolytic bacterial counts / ml	Not to exceed 5
Thermophilic bacterial count / ml	- 0 -
Slime-forming organisms	- 0 -

Apart from dairy uses, a major portion of water is used for steam generation and cooling, these waters must comply with the requirements specified in Chapters 46 and 47 respectively.

CHAPTER - 62

DISTILLERY

(See also, Chapters – 72 & 55 "MALT PREPARATION" and "BREWERY")

Distillery is an industry where alcohol is produced from fermented liquor by distillation. Alcohol (ethanol) is manufactured from a variety of natural products. Among them, grains, fruits and molasses are very important.

Grains are normally ground and mixed thoroughly with sufficient quantity of water which is known as 'mash'. The conditions* are so adjusted for a better fermentation. Yeast is the best known agent carrying out fermentation which is added to the mash in calculated quantities. After 7 – 10 days, the fermentation will be completed and the fermented liquor is then subjected to distillation whereby alcohol (95%) is obtained. This is the basic procedure followed in a distillery. The type of liquor depends on the grains used for fermentation. For example, making a beer from a grain mixture containing atleast 51% corn and distilling and ageing, yields bourbon whiskey. Similarly rye whiskey must start with 51% rye in the grain mixture to be mashed or fermented. Similarly corn is also used to a larger extent in distilleries to produce beverage alcohol.

Wine is prepared from grapes. The grapes are run through a crusher, which macerates them but does not crush the seeds. The resulting pulp (termed as 'must') is pumped on to fermenting tanks. Then an active culture of selected and cultivated yeast is added. On completion of fermentation, the juice will be converted to wine. The wine may be subjected to some other processes to render it clear and to improve taste and may be used directly as a liquor. Or the fermented wine may be subjected to distillation to obtain brandy.

Most of the industrial alcohol is derived from molasses. Molasses is the dark brown liquor obtained on separation of sugar crystals from the mother liquor of processed cane juice at cane sugar industries. It is a viscous liquid containing 50% fermentable carbohydrates. It is diluted first to have a sugar concentration of about 15% - 17% as yeast cannot ferment the concentrated molasses. The diluted molasses is termed as 'mash' and is pumped into fermentors. The pH of the mash is adjusted to lie between 4.0 and 4.5 to suppress bacterial action and multiplication of wild yeasts. Then selected culture of yeast in inoculated into the

* Conditions include temperature control, addition of certain chemicals that influence the growth of yeast etc.,

mash for fermentation. To satisfy the nutritional requirements, sometimes ammonium salts, magnesium sulfate, super- phosphate and urea are added. Fermentation essentially completes at the end of 36 – 50 hours. After fermentation period the liquor in the fermentor is distilled for the recovery of alcohol. Ethyl alcohol is the major raw material for the production of several organic compounds and hence the ethanol produced from molasses in the distilleries is sent to other chemical manufacturing industries. It is also used for beverage making.

Distillery is a water intensive industry. Enormous quantities of water are required in a distillery. About 60 – 100 litres of water is required for the production of 1 litre of alcohol. The water used in the distillery should be colorless and clear. It should not have any undesirable odor or taste. Freedom from iron and manganese is essentially required. Though hard waters are undesirable in a distillery, some hardness (150 – 250 mg/l as $CaCO_3$) is usually desirable in the water used for mashing. Moderate amounts of carbonate hardness (50 – 100 mg/l as $CaCO_3$) are required as it acts as a buffer in maintaining the proper pH value during fermentation. The water should not contain any residual chlorine because it destroys flavoring constituents. Presence of organic matter in water is highly undesirable as it is likely to cause putrefaction during fermentation of the grain. This in combination with calcium sulfate may produce hydrogen sulfide, which would impart an offensive taste and odor to the distilled spirit.

The water should be of approved bacteriological quality. Total absence of coliform organisms is an essential requisite. (The limit for MPN of coliform is zero per 100 ml). Coliform organisms are the best lactose fermenters and the lactic acid produced by lactose fermentation would diminish the yeast activity. The number of other bacteriological organisms as tested by Standard Plate Count method should not exceed 100 per ml.

The water used for deproofing* should be absolutely colorless and clear. It should also be free from any odor and taste. The freedom from taste and odor is especially important, for even traces of taste and odor may seriously impair the quality of the product. Similarly the water should also be free from mineral matter. Calcium and magnesium present in water would be precipitated by the alcohol causing cloudiness. Hence only demineralized water has to be used for deproofing. Besides distillery uses, a major portion of water is used for steam generation and cooling. To know the quality requirements of water for these operations, refer Chapter – 47 "BOILER WATER REQUIREMENTS" and Chapter - 48 "COOLING WATER REQUIREMENTS".

---

* Deproofing is a term used in distillery which indicates the dilution of alcohol to the required extent. The ordinary alcohol of 95% strength is 190 – proof alcohol. Absolute alcohol (pure, anhydrous) of 99.8% strength is 200 – proof alcohol.

CHAPTER - 63

EDIBLE OIL REFINERY

In edible oil refineries, cooking oil and hydrogenated oil are the major items produced. Soap is produced as a by product. Some refineries also manufacture soap powders, washing powders and liquid soaps.

In an edible oil refinery vegetable oils are subjected to three types of treatment, namely, alkali refining, bleaching and deodorization for refinement.

The alkali refining is the process in which the edible oils are treated with caustic soda solution to remove the excess free fatty acids. The free fatty acids react with the alkali and yield soap which is separated by settling. Huge amounts of soap is produced in such a way and is sold out. In order to keep the quality of the soap it is desirable to use only softened water to prepare caustic soda solution. In the process of bleaching, color of the oil is removed by passing through a column containing a special type of adsorbent clay. In this process no water is used. Odor of the oil is removed in the third process. Deodorization by blowing superheated steam (200° – 260°C) through the oil under a high vacuum. This steam is condensed and cooled in spray pond.

(For salad oils, prior to deodorization, the bleached oil is subjected to winterizing-a treatment that consists of cooling the oil to a low temperature at about 5°C and filtering the crystallized stearin. Shortening is another process in which bleached oil is hydrogenated and bleached again before deodorization. These processes do not require any water except for cooling.)

To summarize, in an edible oil refinery, water is used principally for steam generation and for cooling. Only a minimum quantity of water is used for processing. The process water used should be colorless and clear, odorless, free from iron and manganese and of approved bacteriological quality. The boiler water and cooling water should meet the requirements specified in Chapters 46 and 47.

CHAPTER - 64

ELECTRONICS INDUSTRY

(Semi Conductor Manufacture)

For the manufacture of semi conductors, printed circuits, transistors and integrated circuits and washing cathode-ray tubes, water with the highest degree of purity is required. Hence ultra-pure water is the only choice. The proper functioning and long term stability of electronic devices could be improved by intensive cleaning (decontamination) of the active crystal surfaces. Ultra-pure water* as a solvent removes from the crystal the alkali and heavy metal ions and other impurities present in the crystal that affects its functioning. Production of ultra-pure water and storage are very tedious and expensive. Virtually all the unit operations of water treatment, namely, chlorination, filtration, softening, carbon filtration, reverse osmosis, deionization, ultra filtration, micro filtration, ozonation, vacuum degasification and ultra-violet sterilization are commonly employed in the production of ultra-pure water. Besides, the design and operation of the storage and distribution systems are also extremely critical. Despite such hardships ultra-pure water proves to be the excellent solvent and more useful in the electronics industry. For more information refer Chapter – 94, "Ultra - pure Water".

The most important reference standards are those of the American Society for Testing and Materials (ASTM) and Semi Conductor Equipment and Materials International (SEMI). These water standards are given below.

Four types of electronic grade water are described in the 1990 Guide by ASTM.

* Ultra-pure water may be considered as a solvent having a purity of > 99.999 percent. It is proved to be a solvent with many physical and chemical advantages.

TABLE – 64.1
ASTM D-19 STANDARD GUIDE FOR ELECTRONIC GRADE WATER D5127-90 (1990)

	Type E-I	Type E-II	Type E-III	Type E-IV
Resistivity, megohm/cm minimum	18 (95% of time) no less than 17	17.5 (90% of time) no less than 16	12	0.5
SiO_2 (total maximum, µg/L)	5	10	50	1000
Particle count per mL	1	3	10	100
Particle size limit, micron	0.1	0.5	1.0	10
Viable bacteria, maximum	1 / 1000 mL	10 / 1000 mL	10/mL	100/mL
Total organic carbon, maximum µg/L	25	50	300	1000
Endotoxins[1], EU/mL	0.03	0.25	N/A	N/A
Copper, maximum, µg/L	1	1	2	500
Chloride, maximum, µg/L	1	1	10	1000
Nickel, mg/L	0.1	1	2	500
Nitrate, maximum, mg/L	1	1	5	500
Phosphate, maximum, mg/L	1	1	5	500
Potassium, maximum, µg/L	2	2	5	500
Sodium, maximum, µg/L	0.5	1	5	1000
Sulfate, maximum, mg/L	1	1	5	500
Zinc, maximum, µg/L	0.5	1	5	500

Note 1- Substances or by-products usually produced by gram-negative micro-organisms which gave a positive test for pyrogens.

Type E-1- Electronic Grade Water. This water will be classified as microelectronic water to be used in the production of devices having line widths below 1.0 micrometer. It is intended that this be the water of ultimate practical purity produced in large volumes and for the most critical uses.

Type E-II- Electronic Grade Water. This water may be classified as microelectronic water to be used in the production of devices having dimensions below 5.0 micrometers. This water should be adequate for producing most high-volume products which have dimensions above 1.0 micrometer and below 5.0 micrometers.

Type E-III- Electronics Grade Water. This grade of water may be classified as macro electronic water to be used in the production of devices having dimensions larger than 5.0 micrometers. This grade may be used to produce larger components and some small components not affected by trace amounts of impurities.

Type E-IV – Electronics Grade Water. Electronics Grade water may be classified as electroplating water for non-critical use and other general applications where the water is in constant contact with the atmosphere because of tank storage.

TABLE – 64.2

SEMICONDUCTOR EQUIPMENT AND MATERIALS INTERNATIONAL (SEMI)

SUGGESTED GUIDELINES FOR PURE WATER FOR SEMICONDUCTOR PROCESSING (1989)

This table summarizes the suggested guidelines* for pure water utilized in semiconductor processing.

Test Parameter	Attainable	Acceptable	Alert	Critical
Residue, ppm	0.1	0.3	0.3	0.5
Total Organic Carbon (TOC), ppm	0.020	0.050	0.100	0.400
Particulates, counts/liter	500	1000	2500	5000
Bacteria, counts/liter	0	6	10	50
Dissolved Silica (SiO_2), ppb	3	5	10	40
Resistivity, megohm/cm	18.3	17.9	17.5	17
Cations, ppb				
Aluminium (Al)	0.2	2.0	5.0	See Note
Ammonium (NH_4)	0.3	0.3	0.5	See Note
Chromium (Cr)	0.02	0.1	0.5	See Note
Copper (Cu)	0.002	0.1	0.5	See Note
Iron (Fe)	0.02	0.1	0.2	See Note
Manganese (Mn)	0.05	0.5	0.1	See Note
Potassium (K)	0.1	0.3	0.1	4.0
Sodium (Na)	0.05	0.2	0.1	5.0
Zinc (Zn)	0.02	0.1	0.5	See Note
Anions, ppb				
Bromide (Br)	0.1	0.1	0.3	See Note
Chloride (Cl)	0.05	0.2	0.8	See Note
Nitrite (NO_2)	0.05	0.1	0.3	See Note
Nitrate (NO_3)	0.1	0.1	0.5	See Note
Phosphate (PO_4)	0.2	0.2	0.3	See Note
Sulfate (SO_4)	0.05	0.3	1.0	See Note

Note: Values not assignable at this time

* In using these guidelines, it should be recognized that some values may depend on the testing method and the calibration techniques. Consequently, the progression and interrelation of values may be more meaningful than the absolute values.

CHAPTER - 65

ELECTROPLATING AND METAL FINISHING

Electro plating industry is also a water intensive industry. Water is required for four main processes, namely,

1. *Cleaning of the material to be electroplated*
2. *Rinsing after cleaning*
3. *Plating bath preparation, and*
4. *Rinsing after plating*

Cleaning is normally carried out using organic solvents and / or alkaline solutions. These alkaline solutions are used singly or in combination. The commonly used alkaline solutions are soap, soda ash, caustic soda, ammonium chloride, tri-sodium phosphate, tetra-sodium pyrophosphate and sodium metasilicate. Hardness of water may be detrimental when soaps or alkaline cleaning agents are used. Hard waters react with soaps / alkaline cleaning agents and produce precipitates resulting in considerable waste of the cleaning solutions. Besides wastage, some of the precipitates formed may adhere to the metal and if they are not removed properly by rinsing, would cause pits in the plating. In the final rinsing of the metal after plating, hard water is reported to cause unsightly streaks. Use of demineralized water will eliminate all such problems.

For rinsing and for plating, water quality is frequently a major problem. High quality water is of primary importance to assure satisfactory finished work. For decorative plating, water spots and stains may necessitate reworking, wiping, buffing, and other laborious operation. Before the application of any organic corrosion-resistant coating, it is almost a necessity to use demineralized water in final rinse, in order to achieve adhesion of the coating. A high concentration of dissolved solids is especially detrimental in rinse waters.

Hard waters should never be used for the makeup of plating baths. Calcium and magnesium are especially troublesome in plating baths, for they tend to precipitate to form scale on the heated surface or a sludge in the water. There is a probability of these precipitates being included in the electrodeposits, causing pits, porosity, and roughness. Magnesium may also reduce the "throwing power" in chromium baths, but on the other hand, magnesium sulfate is sometimes added to nickel-plating baths to produce softer deposits, to minimize certain types of pitting, and to increase throwing power. In acid plating baths, where sulfuric acid and sulfate is used, concentrations of the bath may lead to the formation of calcium sulfate scale, if hard water is used for making up the bath. Hence demineralized water only has to be used for making up the plating baths.

Sodium and potassium are generally not harmful in plating operations, although sodium may cause brittle deposits in nickel baths or reduce the throwing power in chromium solutions.

Iron is one of the most troublesome elements in many plating operations. In a nickel-sulfate bath, it may cause hazy, streaked, pitted, or brittle deposits; in acid copper solutions it produces rough deposits; in chromium baths it reduces the throwing power; in cadmium cyanide it causes hazy deposits; in silver cyanide stained deposits; and in zinc sulfate baths, it lowers the plating efficiency and the protective value of the coating.

Among the anions, bicarbonates are detrimental in heated alkaline baths, for they tend to be converted to carbonates and accelerate the precipitation of calcium. Moreover, they buffer the water and require higher dosages of acid or alkali to obtain the desired pH value. Chlorides have been reported to cause rough, modular, irridescent, and crystalline deposits in cadmium, copper, silver, and tin baths respectively. Chlorides are also harmful in zinc-cyanide plating baths. Chlorides, sulfates and carbonates are harmful in silver plating too. Therefore to prepare plating baths, demineralized water only has to be used. In aluminium anodizing too, demineralized water is the right choice as chlorides and sulfates may build up over the surface. Organic substances reduce chromium, and cause rough, hazy, streaked, colored, or pitted deposits under various conditions. Color and turbidity are similarly objectionable.

Free carbon dioxide in water is also undesirable as it is reported to cause embrittling and edge cracking in nickel plating. In silver planting also, water with a very low content (below 15 mg/l) of carbon dioxide is required.

CHAPTER - 66

FERMENTATION INDUSTRIES

Fermentation is a natural phenomenon carried out by micro organisms. Using fermentation for the production of alcohol from fruits was even known to primitive humans. This phenomenon has been cleverly used by subsequent generations and now it has reached newer heights that it is an important technic for the production of various materials starting from beer to antibiotics.

The commonly employed micro organisms are bacteria, yeast and fungi. Given below is the short list of materials that are produced by fermentation.

1) Beverages	*: Beer, brandy, whiskey, rum, gin and wine.*
2) Alcohols	*: Ethanol, butyl alcohol.*
3) Ketones	*: Acetone and dihydroxy acetone.*
4) Organic Acids	*: Citric acid, Gluconic acid, Lactic acid, Acetic acid, Itaconic acid, Kojic acid and Fumaric acid.*
5) Pharmaceutical products	*: Antibiotics (Penicillin, Streptomycin, Chlorampenicol, Erythromycin and the Tetracycline group and a wide variety of other antibiotics). Vitamins (Riboflavin (Vitamin B2), Vitamin B12) Amino acids Steroids.*
6) Enzymes	*: Proteases, Carbohydrases, Nucleases, Oxidases and Hydrolytic enzymes*
7) Other Substances	*: Glycerin, Sorbose, 2-Keto gluconic acid, 2,3 – Butanediol.*

Water is extensively used in fermentation industries both as a raw material and for other uses such as steam generation, cooling, washing etc., The quality criteria of water vary from industry to industry. However the following generalization is made.

The water used in fermentation industries should be colorless and clear, free from sediments, tastes and odor. As regards mineral matter content, the water should conform to potable water quality. Certain mineral constituents such as calcium and magnesium bicarbonates in water may be advantageous in furnishing additional nutrients for the yeast.

Presence of free chlorine in water intended for fermentation industries especially breweries and distilleries is highly objectionable as residual chlorine destroys some of the organic materials used for flavoring. Hence it has to be removed before proceeding further by passing the water through activated carbon filter.

Freedom from heavy metals and toxic substances such as arsenic, cadmium, chromium, copper, lead, mercury, selenium and cyanide is essential.

The water should be of approved bacteriological quality (i.e) the Standard Plate Count should not exceed 100 per ml and total absence of coliform bacteria (MPN index zero per 100 ml) is required. Coliform organisms are the best known lactose fermenters and commonly found in water. They ferment lactose to lactic acid which diminishes the activity of the yeast leading to a very poor fermentation.

Also refer the following Chapters for more information.

1) *Chapter – 55 Brewery*
2) *Chapter - 72 Malt Preparation*
3) *Chapter - 62 Distillery*
4) *Chapter - 78 Pharmaceutical Industry*
5) *Chapter - 47 Boiler Water Requirements*
6) *Chapter - 48 Cooling Water Requirements*
7) *Chapter - 54 Bottle / Utensil / Food Equipment Washing*

CHAPTER - 67
FERTILIZER INDUSTRY

Fertilizer is one of the largest industries. Though it requires very large quantities of water it is to be mentioned that water is mainly used for steam generation and cooling.

Fertilizer industry can be conveniently classified into three groups and studied.

I. FERTILIZER RAW MATERIALS

The raw materials, phosphorous, potash and sulfur are obtained by mining for which no water (or little water) is required.

II. FERTILIZER INTERMEDIATE

1) Sulfuric acid 2) Phosphoric acid 3) Nitric acid and 4) Hydrochloric acid. These are manufactured in the fertilizer industry itself. For quality requirements of water used for production of these acids refer Chapter – 57 "CHEMICAL INDUSTRIES"

III. FERTILIZER PRODUCTS

The fertilizer products are classified as Nitrogen fertilizers, Phosphorous fertilizers, Nitrogen – Phosphate fertilizers, NPK fertilizer and Blended fertilizers.

For the manufacture of Nitrogen fertilizers, ammonia is the chief raw material and its production is the major process in a fertilizer industry. It is synthesized using nitrogen and hydrogen. The required nitrogen is obtained from the atmospheric air. Hydrogen is obtained from any one of the following: Natural gas, Naphtha, Fuel oil, Coke-oven gas, Coal and electrolysis of water. Steam reforming of natural gas is the most widely employed method for ammonia production. The other methods are 1) Partial oxidation process, 2) Coal-gasification process and 3) Combustion of hydrogen-nitrogen mixtures.

For all these processes, steam is essentially required than water. Large quantities of water are being employed for the production of steam and cooling purposes. For quality requirement of boiler feed water and cooling water refer Chapters - 47 and 48.

Urea, ammonium sulfate, ammonium chloride, ammonium nitrate and calcium-ammonium nitrate are produced using ammonia. The water required for the production of these fertilizers shall meet the requirements for a water used in chemical industry. See Chapter – 57.

The two important phosphatic fertilizers are Normal Superphosphate [$Ca_2H_2(PO_4)_2$ + $CaSO_4$], and Triple Superphosphate [$Ca_2H_2(PO_4)_2$]. To obtain normal superphosphate, well ground phosphate rock and sulfuric acid (60 - 70%) are mixed rapidly. Triple superphosphate is manufactured by treating ground phosphate rock with phosphoric acid. The phosphoric acid is normally produced in the fertilizer industry itself by treating ground phosphate rock either with sulfuric acid or nitric acid or hydrochloric acid.

The two important nitrogen – phosphate fertilizers are Mono Ammonium- Phosphate (MAP) [$NH_4H_2PO_4$] and Diammonium Phosphate (DAP) [$(NH_4)_2HPO_4$]. Ammonia and phosphoric acid (30 – 40% P_2O_5) are allowed to react to obtain mono ammonium phosphate. Diammonium Phosphate is produced by reaction of mono ammonium phosphate with ammonia.

NPK fertilizers are produced in three ways. Dried fertilizers (nitrogen, phosphate and potash fertilizers) are blended in suitable proportions. Hence no water is required for this process. In the second method, potash particles are coated with diammonium phosphate. The third process is identical with phosphoric acid acidulation, but nitric acid is used in place of phosphoric acid and at later stages potassium chloride is added.

The quality requirements of water for nitrogen fertilizers, phosphatic fertilizers, nitrogen-phosphate fertilizers and NPK fertilizers are similar to that of water used in chemical industry. Refer Chapter - 57 "CHEMICAL INDUSTRIES".

CHAPTER - 68

FOOD INDUSTRY

Food industry covers a variety of food manufacturing industries encompassing multifarious operations. It is one of the industries where huge volumes of water are used. The water use in these industries may be broadly classified into two categories, namely, Food Processing and Food Canning which are discussed below.

FOOD PROCESSING

In food processing, water is used for a variety of purposes besides processing, such as for boiler–feed, cooling, washing, flushing and other general purposes.

Process water uses include washing of raw materials and process equipment, dissolving or extracting and addition of required materials to the product and conveying the materials from one area to another.

Steam is generated in a food industry for cooking, for heating evaporators (or sometimes space heating). In some industries power is generated in which case, first the steam is used to rotate turbines and then sent for processing. If steam comes into direct contact with food, there are strict limits on chemicals used for water treatment which should be strictly adhered to.

Cooling water in food industry is used to cool process equipments such as compressors, cookers and engine jackets, to operate refrigeration equipment and to condense steam etc.,

Since water is incorporated in many of the products, the various substances dissolved in water enter the physical and chemical reactions. Certain constituents of water are advantageous and desirable, and others may be detrimental. Similarly the presence of some of the mineral matter is advantageous to certain end products, while the same mineral matter is undesirable for some other products. For example, water high in sulfate hardness may be excellent for certain fermentation processes but unusable for canning peas, beans or lentils

Owing to the great number of foods produced and the complexity of the reactions, a generalization of water quality requirements or threshold concentrations of constituents cannot be made. The threshold concentrations will differ for the various kinds of foods being processed. The water should atleast meet the requirements of a drinking water, is one requirement commonly recommended for nearly all types of food processes. Generally the water should be clear, colorless, free from iron and manganese and from objectionable tastes and odors. The hardness should be low. The water should be of good bacteriological quality*. In many cases, because of the danger of spoilage or after-growth, precautions must be taken against the presence of micro-organisms other than those normally considered to be pathogenic to humans. It is recommended to practice ozonation for the disinfection of water as chlorination may impart undesirable tastes and odors to water which will be ultimately passed on to the food product.

The other common requirement irrespective of the food product manufactured is freedom from iron, manganese (either alone or sum of both not to exceed 0.2 mg/l), color, turbidity, odor and objectionable tastes. For specific requirements of individual food industries, readers are directed to refer the following chapters.

Chapter - 53 Beverage Production
Chapter - 58 Cider Making
Chapter - 59 Confectionary
Chapter - 61 Dairy
Chapter - 63 Edible oil Refinery
Chapter - 70 Ice manufacture
Chapter - 73 Meat packing
Chapter - 83 Soft Drinks manufacture
Chapter - 84 Starch manufacture
Chapter - 86 Sugar industry
Chapter - 93 Miscellaneous Industrial uses

Separate water quality criteria for washing is set by various agencies. It is to be mentioned that since washing of food equipment is an important operation in food processing

* Total absence of coliform organisms is required. The Standard Plate Count should not exceed 50 per ml.

the quality requirements for water used for washing and more stringent. For more details refer Chapter – 54 "BOTTLE/ UTENSIL / FOOD EQUIPMENT WASHING"

FOOD CANNING

A primary consideration for the quality of water used in food canning and especially in freezing processes is its bacterial content. The water should be of approved bacteriological quality. Not only must such water be free of pathogens that maybe transmitted to the consumer, but it must also be sterile with respect to saprophytic organisms that may cause spoilage.

As regards the physico-chemical characteristics, water used for canning should be colorless and clear, free from iron and manganese and objectionable tastes and odors. Compounds of iron, manganese, copper and organic matter in water are reported to cause undesirable color, odor and taste in canned food. Hence total elimination of these pollutants is the major requisite for a water intended for canning. This apart, any substance that may likely to deteriorate the quality of food[*], decrease vitamin content[†], toughen the food and cause deposition should also be avoided.

Hardness is also undesirable as it causes toughening of some vegetables. Low hardness water is preferable for canning peas, beans, and lentils as hardness combines with some of the protein making the product harder and imparts a mealy taste. The turbid juice often found in canned peas is not due to calcium salts precipitated from the water, but due to starchy matter. It is also reported that calcium reacts with oxalates in the beet root to form white deposits of calcium oxalate, which are normally noticeable on cut beets. As for fruits are concerned, hardness is reported to have nil or no effect. Rather it is beneficial to vegetables and fruits that tend to soften in canning–string beans, peaches, pears etc., – because it hardens the product making it crisp and retentive of its shape. (calcium sulfate toughens the vegetable).

[*] For example, copper is reported to decrease the shelf life of packaged foods, especially dairy and vegetable oil products.

[†] Peas and snap beans that had been blanched with waters containing various concentrations of calcium and magnesium showed a greater loss of ascorbic acid (vitamin-C).

It is reported that even softened waters that contain high alkalinity is used for sterilizing, the outside of the cans present a spangled appearance. In such cases, the alkalinity has to be lowered to about 50 mg/l. The presence of fluoride is also undesirable as it tends to concentrate in spaghetti, baked beans and cereals. The various threshold or limiting values that have been recommended are summarized below.

TABLE – 68.1

RANGE OF RECOMMENDED THRESHOLD VALUES FOR THE CONSTITUENTS IN WATER USED FOR FOOD CANNING*

Constituent	Range of Recommended Threshold values in mg/l
Total dissolved solids	850
Turbidity	1.0 - 10
Taste and odor	None - low
Hardness (as $CaCO_3$)	50 - 85 (general) 200 - 400 (peas) 100 - 200 (fruits & vegetables) 25 - 75 (legumes)
Iron	0.2
Manganese	0.2
Iron and manganese	0.2 - 0.3
Sodium chloride	1000 - 1500
pH value	Not lower than 7.5
Hydrogen sulfide	1.0
Fluoride	1.0
Alkalinity (as $CaCO_3$)	30 - 250
Nitrates (as N_2O_5)	15
Ammonia (as NH_3)	0.5

* Reproduced with the kind permission of the Resources Agency of California, State Water Resources Control Board, California from their publication "WATER QUALITY CRITERIA", 1963, edited by J.E. Mckee and H.W. Wolf.

CHAPTER - 69

HUMIDIFICATION

Like other processes, a bulk of water is also utilized for humidification in industries, especially in textile mills. The water used for humidification should be clear and free from suspended impurities. As water is sprayed through nozzles (usually with very small apertures), any slight amount of suspended matter present in water will clog the pores. The water should not have the tendency to deposit scale. Generally softened water is used for humidification. The softening has to be carried out by ion-exchange method only. Softening by the addition of chemicals is not recommended.

The water should not be corrosive and contain nuisance organisms such as iron bacteria, the filaments of which will clog the pores in course of time. Suitable treatment such as chlorination has to be practised to eradicate the nuisance organisms from water.

CHAPTER - 70

ICE MANUFACTURE

Ice is manufactured in a number of food industries and in cold storage plants as well as in separate units. As the manufactured ice should be clear, colorless and tasteless, the water used to make it should have similar qualities. A good ice should be free from bubbles, snowy butts and heavy cores as for possible. Also it should not shatter when handled.

The water used for making ice should be scrupulously potable both from the point of view of its chemical composition and its bacteriological condition. Since ice is a home for microbes, the water used for making ice should be absolutely free from bacteria. Especially, absence of coliform organisms is an essential requisite.

The water should also be free from iron and manganese and not too high in mineral matter content. In freezing large blocks of ice, the dissolved mineral matter tends to concentrate in the central core, making it cloudy. If iron and manganese and organic matter are present, they result in a colored core and an unattractive ice cake. Calcium and magnesium cause opacity of the ice. The carbonate hardness is very objectionable, as both calcium carbonate and magnesium carbonate are largely precipitated on freezing forming slimy and gritty deposits (Preferably the carbonate hardness of the water should be under 70 mg/l). Sodium bicarbonate is especially undesirable as it causes white butts, heavy cores and retard freezing. Besides it may cause shattering. The chlorides and sulfates of calcium and magnesium are also undesirable. If they are present in large amounts they act like sodium salts (i.e.) the concentrate salt solutions require much lower temperature to solidify. Sodium sulfate is also reported to cause a white shell in ice.

Fluoride in water is reported to cause serious brittleness and cracking of ice blocks when its concentration approaches 1.5 mg/l. Absence of heavy metals in water is also desirable. If high-grade ice required it has to be prepared from distilled water / demineralized water / RO treated water.

CHAPTER - 71
LAUNDRY

Water intended for laundry uses should be free from color, turbidity, iron, manganese and aluminium. The water should be soft – an important criterion.

The most important constituent to be avoided or eliminated in water used for laundering is hardness, because of the wastage of soap and the formation of curds. (It is well to remark, however that hard water can be used in rinsing and blueing the goods). Also of great detriment is the presence of iron and manganese which, even in low concentrations, can cause stains on the materials being laundered. Manganese is especially objectionable in water used in laundry work and textile processing. Concentrations as low as 0.2 mg/l may cause dark brown or black stains on fabrics. High bicarbonate alkalinity, too may cause staining difficulties. A peculiar phenomenon is observed with waters containing sodium bicarbonate. Sodium bicarbonate has an yellowing effect on clothes laundered with water containing it. This is due to the formation of an oxycellulose when the bicarbonate concentrates under the heat of the iron. This effect is marked when the bicarbonate alkalinity exceeds 200 mg/l. (as $CaCO_3$). It is noticeable above 100 mg/l, but is unobjectionable below 60 mg/l. Needless to say that the water should be free from color and turbidity which may also leave colored spots or impart the white clothes a dull appearance.

A summary of the range of recommended threshold and limiting concentrations of various constituents in waters used for laundering processes is presented.

TABLE – 71.1
RANGE OF RECOMMENDED THRESHOLD AND LIMITING CONCENTRATIONS FOR LAUNDERING*

Constituent	Range of Recommended Threshold and limiting Concentrations in mg/l
Hardness, as $CaCO_3$	0 – 50
Iron and manganese	0.2 - 1.0
Iron	0.2 – 1.0
Manganese	0.2
Alkalinity(as $CaCO_3$)	6.0
pH value	6.0 - 6.8

* Reproduced with the kind permission of the Resources Agency of California, State Water Resources Control Board, California from their publication "WATER QUALITY CRITERIA", 1963, edited by J.E. Mckee and H.W. Wolf.

CHAPTER - 72

MALT PREPARATION

Malt is prepared from grains. Barley is the most widely employed grain. It is malted to develop necessary enzymes and desired flavor. Malting is a simple process in which grains are steeped in cold water and spread out on floors or in special compartments for germination. At the proper time, (5 – 8 days) when necessary enzymes are formed, green malt is placed on driers and dried to a predetermined moisture content and stored.

In the malt preparation, water is used primarily in soaking the grains. The water used for soaking grains should be relatively free from microorganisms. Steeping solution gradually becomes an excellent medium for the growth of bacteria and even a few number of microorganisms present in water multiply rapidly and deteriorate the product in the initial stages itself.

During steeping, soft waters extract from the grain more mineral and organic substances (soluble protein derivatives) which are valuable nutrients for yeast. On the other hand, hard waters extract more albuminoid matter. Hence a moderately hard water is preferred. But the presence of magnesium carbonate is undesirable as it is harmful to the enzymes amylase. Both calcium and magnesium carbonates are reported to retard germination.

Other quality requirements of water used to steep grains for malt production are described below:

The water should be free from objectionable tastes and odors. A low content of nitrate is preferable. Similarly freedom from iron, and ammonium salts is recommended. The barley is colored brown by iron salts if present in water. The salts of iron have also a harmful effect in as much as they retard fermentation. This apart, iron if present as oxide would lessen the activity of the enzyme diastase of the malt by forming an insoluble compound with it. Water containing sodium chloride and calcium chloride is also undesirable as they interfere with the germination of grains. The presence of organic matter is highly undesirable. When the grain is steeped in water containing organic matter it provokes a putrid fermentation of the grains and make them unfit for malting, or the tendency to mouldiness is favored with resultant defective malting.

CHAPTER - 73

MEAT-PACKING

(See also, Chapter - 68 "FOOD INDUSTRY")

Meat packing is a major industry which covers a variety of main products and byproducts. Huge quantities of water is used for canning and other processing operations, and for washing, cleaning, flushing, boiler feed and cooling, besides general use.

Water intended for meat packing operations should be colorless and clear, odorless, tasteless, free from iron, manganese and organic matter and of approved bacteriological quality. Though hard water does not appear to be injurious to the curing and preparation of meat for consumption, it is reported that a better color is obtained with soft water than with hard water. For flushing and cleaning operations, the above quality is sufficient; no more treatment is required.

A special mention is made about nitrates. In the production of cold meats and sausages, the presence of nitrate in the process water is reported to cause discoloration of the final product. Therefore it is essential to remove the nitrates in process water either by ion exchange method or by reverse osmosis.

CHAPTER - 74
MINING

From the earliest days of mining, water has been the main agent for separating the precious metals and the more valuable ores from their gangue, and it still remains such whenever an ore is associated with a large proportion of foreign matter.

For the concentration of ores by floatation, the water used should be alkaline. Alkaline waters facilitate the settling of pulp and slime. Waters are rendered alkaline by the addition of calcium oxide or calcium hydroxide or soda ash to pH 9.0 to 9.5. Soda ash is preferable, as lime tends to deposit (as $CaCO_3$) on the particles of ore that would interfere with their wetting by the floatation chemicals. At gold mines using cyanide process, lime is used as it is the cheapest chemical.

As a general rule it may be stated that waters used in the mineral industry should be free from organic matter and of not too high in mineral matter content.

CHAPTER - 75

NUCLEAR POWER PLANTS

(See also, Chapter - 91 "THERMAL POWER PLANTS")

Nuclear power plants contribute significantly to power generation and several nuclear power plants have been commissioned across the world today. In a nuclear plant, the heat required to convert water to steam is obtained by nuclear reaction. In the plant, heat is generated from the fission of heavy nuclides in contrast to conventional power generating plants where heat is generated by the combustion of fossil-fuels. Hence, machineries, equipments and other appurtenances of both of these two are entirely different until heat generation section and thereafter, the two types of plants are essentially the same. These power plants require water of very high purity.

Nuclear power plants employ two types of reactors, namely,

1) *Pressurized Water Reactors (PWR), and*
2) *Boiling Water Reactors (BWR)*

In pressurized water reactor, heat generated in the nuclear core is taken up by water (reactor coolant) circulating at high pressure through the primary circuit. The absorbed heat is transferred to the secondary system in a heat exchanger or boiler, thereby generating steam in the secondary system. (Details given in the next section).

In boiling water reactor, steam is generated in the core and delivered directly to the steam turbine.

PRESSURIZED WATER REACTORS (PWR)

Pressurized water reactor has two major water systems.

1) *The Primary Loop or Reactor Coolant System*
2) *The Secondary Loop or Steam Generator–Turbine Cycle.*

In the primary loop, the coolant or core heat-removing medium is water. The coolant water also serves as a moderator. High purity water is an essential requisite for primary loop.

The purity must be of high order so as to preclude fouling of heat-transfer surfaces, corrosion and the formation of radio-active materials through nuclear activation processes. Needless to say that corrosion if any occurs, would impair the material integrity or results in the increased concentration of corrosion products that may lead to severe consequences. Therefore superior quality water is essential in the nuclear sector of the cycle. The quality of water must be similar to that of the effluent of a mixed bed ion exchange demineralization. Infact, ultra pure water is the most preferred one. The quality requirements of makeup water is tabulated below. (Table 75 – 1)

Table 75 – 1
Reactor – Coolant Makeup Water – Quality Requirements

S.No.	Parameter	Limits
1.	Conductivity μSiemens / cm	< 0.5
2.	Total solids, ppm	< 0.5
3.	Chlorides, (as Cl), ppm	Not more than 0.15
4.	Fluorides (as F), ppm	Not more than 0.1

It is reiterated that chloride if present in concentrations more than 0.15 ppm is highly undesirable as it corrodes the austenitic stainless steels to a significant extent. Dissolved oxygen also corrodes the stainless steel when the concentration exceeds 0.10 ppm.

Unlike in the normal treatment of water, the coolant water is added with boric acid. It is added mainly to control the fission reaction. Controlled addition and removal of boric acid in the primary loop water provide the correct concentration of neutron–absorbing boron needed to control neutron flux and energy transfer. This apart, the presence of boric acid fixes the lower pH specification of 4.5. In addition to boric acid, chemicals such as lithium hydroxide, forming relatively safe radio isotopes under neutron flux are also added for pH control in the primary loop. However such additions should also be added in calculated quantities and their concentration should not exceed the following limit.

Additives, ppm .... < 25 of Li[†], K.

Primary loop water purity is usually maintained by continuously passing a portion of the circulating water through a mixed bed demineralizer. In these beds, anion resins in the form of

† El–Wakil, M.M., "Nuclear Power Engineering" – 1962 Mc Graw – Hill Book Co., New York.

borate are used to avoid removal of boron from water. (As impurities accumulate, these resins tend to become ratio–active and regeneration becomes impractical. These resins are then disposed of as a solid rad waste).

The heat generated in the primary loop is absorbed by water and transferred to the secondary system in a heat exchanger or boiler. The water used for steam generation in the secondary loop (or steam generator–turbine cycle) should meet the requirements specified for high pressure boilers. It should be remembered that the primary coolant water (which picks up radioactivity from a number of sources) is separate from the water which serves as working fluid for the steam turbine and the steam generator serves as a barrier between the two fluids.

Two basic heat exchanger designs are employed as secondary loop steam generators in PWR systems. The recirculating U–tube type is analogous to a drum type boiler. The other PWR heat exchanger design provides once through steam generation. Both these steam generators require maximum corrosion resistance on the primary side. Therefore, the tubes are commonly constructed of a special nickel–iron–chromium alloy. The quality requirements of water used for steam generation in both these systems are similar to those of high pressure boilers.

BOILING WATER REACTORS (BWR)

As stated already, in boiling water reactor, steam is generated in the core and delivered directly to the steam turbine. Therefore the construction of BWR is entirely different from that of PWR.

Boiling water reactors are comparable to drum–type boilers. A portion of the water passing through the core is converted to steam. The steam–water mixture is then separated. The steam is sent to turbine section and the separated water is returned to the core inlet. It must be remembered that the same water serves as the cycle working fluid, the reactor coolant and the nuclear reaction moderator.

An important aspect of BWR operation is the effect of direct exposure to nuclear radiation on the working fluid. Besides, the steam produced by a BWR reactor contains high concentrations of oxygen, which is highly corrosive in nature. This apart, nitrogen that enters the cycle by air in-leakage and/or by decomposition of nitrogenous compounds would be converted to nitric acid which is also a major factor of corrosion. Neither oxygen nor nitric acid that is formed can be suppressed by oxygen scavengers and additives for pH control as they are subject to nuclear decomposition. This fact is emphasized as the corrosion control relies primarily on the corrosion – resistant materials used for construction and the material should not be affected by the impurities of water. Therefore every effort is taken to treat water scrupulously so that it should not contain any mineral matter or metallic contaminants and substances that may corrode the construction material. The makeup water is demineralized to an extent that its conductivity should be less than 1.0 µs/cm* and silica content below 10ppb**. The condensate is demineralized and it should also conform to the following limits: 0.1 µs/cm at 25°C, less than 2 ppb copper***, less than 30 ppb metals and concentration of chlorine below 2 ppb. Resultant feed water should be essentially neutral with a pH equal to 7.

Whatever may be the type of reactor (either PWR or BWR), the most important condition regarding water is that it should not contain cobalt (naturally occurring). It is an extremely undesirable constituent of primary coolant water passing through reactor. The product would be highly radio active, emitting high–energy gamma rays during subsequent decay. Hence utmost care must be exercised to avoid cobalt in such systems.

---

* µs/cm – microSiemens / cm – Unit of electrical conductivity

** ppb – parts per billion (micrograms per litre)

*** Copper is specifically limited because of its tendency to foul the orifices of the core water distributor.

CHAPTER - 76

OIL – WELL FLOODING

When an oil reservoir is penetrated, oil and gas are forced to the surface by the natural reservoir pressure. Emerging of oil from the natural reservoir by natural forces may extend from a few months to years. This period is termed as 'Primary Production'. When the natural reservoir energy subsides, some method is employed to bring out the oil. The methods include water flooding, gas injection and other processes involving fluid or energy injection. These methods of extraction are collectively termed as 'Secondary Production'*. Water flooding is the most common method of secondary recovery.

Water flooding is nothing but injecting water into the oil-wells, from a series of drilled wells, known as injection wells. A desirable water for this purpose is that produced from the formation after separation from the oil and gas. However the requirement of water for flooding is greater and the quantum of produced water is not sufficient and hence supplemental water is also injected which is derived from injection wells drilled uniformly throughout the oil-well.

The permeability of the reservoir has a great bearing on the suitability of water for flooding. Hence both the formation rock and the water to be flooded have to analyzed in great detail. Some of the quality requirements for water flooding are given below:

The most important criterion is that the water should be free from turbidity. The turbidity would clog porous strata and create high back pressures. When produced water and supplemental surface waters are used, filtration and/or chlorination may be required to remove residual oil accumulations, sand or dirt, organic and/or inorganic suspended solids. The water should not contain more than 0.1 mg/l iron. The pH should be more than 7. Low alkalinity waters are not preferable for flooding. The water should not be corrosive. Presence of hydrogen sulfide is highly undesirable.

Algal growth may likely to occur in water stored for flooding operations. Such growths should be prevented by the addition of suitable algicides.

* After secondary recovery, still a large portion (about 50%) of the petroleum remains in the reservoir which is extracted by some other methods and this final recovery is called as 'Tertiary Recovery'.

Most of the injection waters are subjected to filtration. The media for filtration and design of the filter units are largely dependent on the reservoir permeability. Normally mixed media beds, or septum filters using diatomaceous earth as a filter aid are employed. Deaeration may be required to reduce corrosion and to prevent iron oxidation.

Another important point to be remembered is that in case if the injection water must include surface water or well water with lower salinity, chemical incompatibility with the formation rock may result. Salinity than the original formation water, chemical incompatibility with the formation rock may result, which may cause swelling of clay in the reservoir rock. Hence to assess the suitability, both the water and formation rock have to be tested simultaneously. The effect of chemical additions and filtration are evaluated from the study of formation rocks.

CHAPTER - 77

PETRO CHEMICAL INDUSTRY

Petroleum refining is one of the largest manufacturing industries. More than 3000 products are derived from petroleum. Solvents, lubricants, chemicals and chemical intermediates, plastics, asphalt are a few to mention. Water in petrochemical industries is mainly used for boiler and for cooling. Besides, a small portion is being used as a medium of transport and also for certain chemical reactions.

Steam is the major source of energy in a refinery. It is used to heat process streams, to strip sour water*, to drive pumps and compressors. Some of the industries generate electricity too. The water used for boiler should meet the quality requirements of boiler water. (for details refer Chapter - 46 "BOILER WATER REQUIREMENTS").

The cooling water requirements of a petrochemical industry is very large. About 80 to 85% of the total water requirements is used only for cooling. The cooling systems in a refinery is similar to that of the cooling system in any other industry. Among them, open recirculating systems are most widely employed. The closed recirculation system is installed for a few requirements. Once through cooling is seldom practised. However petrochemical industries located in coastal areas prefer to install once-through cooling water system due to the availability of sea water for cooling.

The quality of cooling water used in petrochemical industry is similar to that of cooling water used in other industries; the water should be free from suspended matter, iron and manganese, it should not form scales and should not be corrosive. The most important is that it should not form clogging organic growths.

* When steam used in the processes is condensed, hydrocarbons also condense along with steam. After separation of the hydrocarbons, the condensed steam contains oils and hydrogen sulfide, ammonia, mercaptans and hydrogen cyanide. The condensed steam containing these impurities is called as sour water (or foul water). These contaminants are removed by steam stripping - a simple distillation procedure.

CHAPTER - 78

PHARMACEUTICAL INDUSTRY

(See also, Chapters "CHEMICAL INDUSTRIES" and "FERMENTATION INDUSTRIES")

Pharmaceutical industry is the life saving industry and thousands of pharmaceutical products are manufactured now. The pharmaceutical industry employs a greater variety and more complicated steps in its manufacturing processes than any other industry. Divulging all the products and manufacturing processes is beyond the scope of this book. However a short list of medicines and processes is provided for information.

Following is the select list of pharmaceutical groups.

1) Analgesics and antipyretics 2) Anesthetics 3) Antibacterials (Antiseptics) 4) Antibiotics 5) Antihistamines 6) Cardio-vascular drugs 7) Central nervous system stimulants 8) Dermatologicals 9) Diuretics 10) Gastro intestinal drugs 11) Hormones 12) Respiratory agents 13) Sedatives and hypnotics 14) Sulfonamides 15) Tranquilizers 16) Vitamins and nutrients 17) Miscellaneous medicines uncovered in the above groups.

Pharmaceutical industry is nothing but a chemical industry carrying out more complicated processes. Some of the important process by which drugs are manufactured are given below.

1) Alkylation 2) Condensation 3) Catalytic Rearrangement 4) Dechlorination 5) Dehydration 6) Extraction 7) Esterification 8) Halogenation 9) Hydrogenation 10) Methylation 11) Oxidation 12) Reduction 13) Sulfonation

This apart, a great variety of medicines are manufactured by fermentation process too.

In a pharma industry, water is used in multifairous ways. Firstly, it is used in the process in which it becomes an ingredient in the product. Water is the principal vehicle for the preparation of solutions and suspensions for injections. Secondly, it is used for boiler and cooling purposes. Thirdly, it is used for general purposes such as cleaning of containers etc.

In the pharmaceutical industry, greater emphasis is placed upon the purity of the products. Needless to say then that the quality of water is vital for obtaining such high purity products. The production water must be of high standard and of high bacterial purity. It must be sterile (free from living microorganisms) and apyrogenic (free from the fever – producing products of micro organisms). Demineralized water or Reverse Osmosis treated water is the best option. In addition to such treatments, the water has to be further subjected to triple distillation at a temperature of more than 230°C in order to remove pyrogens.

A special mention about copper is made here. It is reported that even minute traces of copper are deleterious in many pharmaceutical preparations. A report by Schou[†] indicates that as little as 0.02×10^{-4} percent of copper in distilled water can affect the stability of many drugs.

The most important pharmacopoeias are the British (BP), European (EP) and the U.S pharmacopoeia (USP). Each pharmacopoeia specifies various grades of water. Among them[*] the specification prescribed by USP for two major types of water namely, **"Purified Water" (PW)**, and **"Water for Injection" (WFI)** are given below.

Purified water has a number of applications (both pharmaceutical and cosmetic) including higher volume uses such as container washing. WFI is used for the preparation of parenteral (injectable) solutions. There is a clear distinction between these two types of water (i.e) in addition to the quality specified for "Purified Water", "Water for Injection" must also meet a bacterial endotoxin (pyrogen) specification.

Excerpts from **USP 23 : Purified Water and WFI Standard** are as follows:

PURIFIED WATER

Category : Pharmaceutic aid (solvent)
Description : Clear, colorless, odorless, liquid

[†] Schou, S.A., "Content of Copper in Distilled Water" – Water Pollution Abstracts, 29 (1956), 408

[*] Six grades of water are used in a pharmaceutical industry. Among them, "Purified Water (PW)", and "Water for Injections (WFI)" are the most important. Both these two are used for the production of remaining four grades of water which inturn are used for the production of bulk pharmaceutical chemicals.

Purified water is obtained by distillation, ion exchange treatment, reverse- osmosis or other suitable process. It is prepared from water complying with the regulations of the United States Public Health Service with respect to drinking water. It contains no added substance.

Note: Purified Water is intended for use as an ingredient in the preparation of compendial dosage forms. Where used for sterile dosage forms other than for parenteral administration, process the article to meet the requirements under Sterility Tests (71), or first render the Purified Water sterile and thereafter protect it from microbial contamination. Do not use Purified Water in preparations intended for parenteral administration. For such purposes use WFI, Bacteriostatic WFI or Sterile WFI.

Packaging and Storage	: Where packaged, preserve in air-tight containers.
Labelling	: Where packaged, label it to indicate the method of preparation.
pH	: Between 5.0 and 7.0, determined potentiometrically in a solution prepared by the addition of 0.30 ml of saturated potassium-chloride solution to 100 ml of Purified Water.
Chloride	: To 100 ml add 5 drops of nitric acid and 1 ml of silver nitrate TS. No opalescence is produced.
Sulfate	: To 100 ml add 1 ml of barium chloride TS. No turbidity is produced.
Ammonia	: To 100 ml, add 2 ml of alkaline mercuric potassium iodide TS. Any yellow color produced immediately is not darker than that of a control containing 30 μg of added NH_3 in High - Purity Water (0.3 ppm).
Calcium	: To 100 ml, add 2 ml of ammonium oxalate TS. No turbidity is produced.
Carbon dioxide	: To 25 ml, add 25 ml of calcium hydroxide TS. The mixture remains clear.
Heavy Metals	: Adjust 40 ml of Purified Water with diluted acetic acid to a pH of 3.0 – 4.0 (using short range pH indicator paper), add 10 ml of freshly

prepared hydrogen sulfide TS, and allow the liquid to stand for 10 minutes. The color of the liquid when viewed downwards over a white surface is not darker than the color of a mixture of 50 ml of the same Purified Water with the same amount of diluted acetic acid as was added to the sample, matched color – comparison tubes being used for the comparison.

Oxidizable Substances : To 100 ml, add 10 ml of 2N sulfuric acid, and heat to boiling. Add 0.1 ml of 0.1 N potassium permanganate and boil for 10 minutes. The pink color does not completely disappear.

Total solids : Evaporate 100 ml on a steam bath to dryness and dry the residue at 105°C for 1 hr. Not more than 1 mg residue remains.

Bacteriological Purity : It complies with the US Public Health Service Regulations for drinking water with respect to bacteriological purity.

WATER FOR INJECTION

Category : Pharmaceutic aid (solvent)
Description : Clear, colorless, odorless, liquid

Water for Injection (WFI) is water purified by distillation or by reverse- osmosis. It contains no added substance.

Note: WFI is intended for use as a solvent for the preparation of parenteral solutions (i.e pharmaceutical solutions injected into the body). For parenteral solutions that are prepared under aseptic conditions and are not sterilized by appropriate filtration or in the final container, first render the WFI sterile and thereafter, protect it from microbial contamination.

Packing and Storge : Where packaged preserve in air-tight containers. It may be stored at a temperature below or above the range in which microbial growth occurs.

Pyrogen : When previously rendered isotonic by the

addition of 900 mg pyrogens – free sodium chloride for each 100 ml, it meets the requirements of the Pyrogen test (It should not contain more than 0.25 USP Endotoxin unit per ml).

Other Requirements **:** It meets the requirements of the tools under Purified Water.

During 1975, the United States Pharmacopoeia has authorized reverse osmosis for the production of Water for Injections. However the product from reverse osmosis has to be end-polished by mixer bed deionization and membrane filtration. Conversely, mixture bed deionized water could be used as the feed water to the reverse osmosis plant. With either method, the process becomes a combination of ion exchange and reverse osmosis. Deionization followed by reverse osmosis with an end filter of 0.2 μm presents a reliable system for the production of Water for Injections.

Storage of pure water is equally important as that of production. Water for the preparation of sterile products should be used as soon as possible after production. It is even more important that infusion fluids, (especially if they contain dextrose or other similar materials) should be sterilized immediately to avoid the growth of pyrogen-producing organisms. Pyrogenecity has been known to develop in as little as one hour after the preparation of such solution.

Several companies offer storage tanks designed to ensure complete protection from chemical or organic contamination. Some of them are equipped with ultra-violet light. Some storage tanks are equipped with heating to maintain a water temperature of about 80^0C to prevent the growth of pyrogen-producing organisms and thus eliminates the risk of the development of pyrogenicity.

WATER FOR HEMODIALYSIS (KIDNEY DIALYSIS)

The limits prescribed by the "Association for the Advancement of Medical Instrumentation" for water used in hemodialysis are given below[*]. The medical concern is to eliminate hemolysis[†] in the blood and the potential for pyrogenic reactions.

[*] For more details, refer the publication, "American National Standards, Hemodialysis Systems".

[†] Disintegration of red blood cells with the liberation of contained hemoglobin.

TABLE – 78.1

WATER QUALITY STANDARD FOR NON-TOXIC SUBSTANCES NORMALLY INCLUDES IN DIALYSATE

Substances	Maximum concentration (mg/l)	Rationale
Calcium	2 (0.1 mEq/l)	Based on allowable margin for error in final dialysate
Magnesium	4 (0.3 mEq/l)	
Potassium	8 (0.2 mEq/l)	
Sodium	70 (3.0 mEq/l)	

TABLE – 78.2

WATER QUALITY STANDARDS FOR SUBSTANCES DOCUMENTED AS TOXIC IN HEMODIALYSIS

Substances	Maximum concentration (mg/l)	Rationale
Aluminium	0.01	Based on the lowest toxic levels reported in the dialysis literature with appropriate margin & safety
Chloramines	0.10	
Copper	0.10	
Fluoride	0.20	
Nitrate	2.0	
Sulfate	100	
Chloride	0.5	
Zinc	0.10	

TABLE – 78.3

WATER QUALITY STANDARDS FOR TOXIC SUBSTANCES REGULATED IN THE EPA DRINKING WATER STANDARDS

Substances	Maximum concentration (mg/l)	Rationale
Arsenic	0.005	10% of EPA level or 'no transfer' level whichever is higher
Barium	0.01	
Cadmium	0.001	
Chromium	0.014	
Lead	0.005	
Mercury	0.0002	
Selenium	0.01	
Silver	0.005	

Since membrane technology has scaled new heights, it is possible to produce waters of very high purity at present at reasonable cost which is a boon to pharma- industry and hospitals.

CHAPTER - 79

PULP AND PAPER INDUSTRY

Paper industry is the most common industry situated all over the world. Production of paper involves two major processes, namely, pulping and paper making. In pulping process, wood and allied materials are digested with chemicals to convert them into pulp. In paper making, pulp is refined and passed over a fine screen to form paper.

PULPING

In the process of pulping, wood or other cellulosic raw materials* are digested with chemicals under high temperature and pressure so that cellulosic fibres of wood are relieved from its binders such as lignin, resin etc., There are three major types of pulping processes employed according to the raw materials and finished products. They are:

SULFATE OR KRAFT PULPING

Employed mainly for the manufacture of strong brown wrapping papers, multiwall papers, bag making papers, gumming papers, paper boards, corrugated boards and strong white papers.

SULFITE PULPING

With this pulping process, fine grade papers such as bond paper, other white grades, tissue paper, book paper and bread wrap paper etc., are manufactured.

SEMI CHEMICAL PULPING

Employed for the manufacture of writing, bond, offset, newsprint and tissue papers and corrugated and specialty boards.

The other pulping processes of less importance are, Soda Pulping, Mechanical Pulping and Chemi – mechanical Pulping.

---

* The cellulosic raw materials commonly used are bamboo, wood, cotton linters, rags, straw mainly of rice and wheat, bagasse, hemp, flax, jute and waste paper.

PAPER MAKING

Based on the type of paper required, the pulps are chosen and disintegrated in a beater, while various fillers* (to fill the pores of paper) and sizes are added to improve the quality of paper. If needed, dyes are also added to impart the required colors to paper. After beating, the pulp is refined. The refining is carried out in a machine, "Jordan" where the fibres are cut to the desired size and the pulp is then passed over a traveling belt of fine wire known as the Fourdrinier wire. Major portion of the water passes through the screen and the fibres form a mat on the surface and is passed on to rollers and calendaring machine for final surface finishing.

The manufacture of pulp and paper is highly dependent upon an abundant supply of high-quality water. The major purposes for which water is used include the digestion or cooking of wood chips to make pulp, repeated washings of the cellulose pulp, transportation of the pulp and other constituents of paper in a thin slurry through preparatory processes and finally onto the paper machine, boiler feed and cooling. Since all the processes are entirely water based, water quality has an important bearing on the final quality of the product. The water used should be free from color, turbidity, suspended solids, iron, manganese and oily mater and without an excessive amount of dissolved solids. Almost all the constituents of water affect the process in one or other way. The impurities and their effects are discussed below.

The most affecting constituent of water is hardness. Particularly the presence of calcium bicarbonate is highly undesirable as it forms scales on or in screens, Fourdrinier wire, deckers, condensers, pumps and recirculating systems (i.e) wherever the water is warmed or evaporated. Hardness also causes difficulties by reacting with sizing compounds and forming precipitates. In processing wood pulp, particularly sulfite paper, insoluble calcium and magnesium resinates are formed; these are deposited on the fibre and make subsequent bleaching difficult. In the soda pulp process, hard waters are objectionable because calcium and magnesium salts are precipitated by the caustic soda. The precipitated calcium carbonate tends to break loose in the form of dissolved chunks, thus contaminating the pulp, and bleaching may not be effective. Magnesium hydroxide does not have this tendency; any magnesium- hydroxide formed tends to disseminate quite evenly in the pulp. In the coloring of

* The commonly used fillers are talc, certain clays, calcium carbonate, blank fixe (barium sulfate) and titanium dioxide. Usually resin or a wax emulsion and aluminium-sulfate (usually referred as 'paper makers' alum') are added to produce a gelatinous film on the fibre and hard surface.

paper also, soft water only should be used or else carbonates precipitate which tend to form lakes with the dyes. Both carbonate and non-carbonate hardness in water tend to increase the ash content, and this is objectionable with certain grades of pulp, especially alpha-cellulose pulps such as those used in the manufacture of rayon and explosives.

When sizing with resin, the mineral salts react with the size, forming insoluble resinates which have no sizing action. The lurking danger with the salts present in water is that they may affect the colloidal condition of the size, tending to make the resin gather into larger particles, thereby reducing the covering power of the size.

Iron, manganese, silica and organic matter are other undesirable constituents of process water. Iron may precipitate and stain the paper. It should not exceed 0.2 mg/l. For high-grade paper in which color is a primary consideration, water containing less than 0.1 mg/l is desirable.

Manganese is more troublesome than iron because of its black oxide which is reported to discolor the pulp fibre and is also detrimental to the stability of resin size. Manganese content of the process water should not exceed 0.05 mg/l.

Turbidity and color can cause trouble in the finer-grade product. Suspended matter is undesirable in process waters because it decreases brightness, affects colors, interferes with texture and uniformity of paper, clogs wire screens and favors growth of slimes. Large quantity of suspended matter is reported to decrease the strength of the paper. If the suspended matter contains sand, it may likely to cause pinholes in paper. Animal matters cause grease spots that will not take color or printing ink.

Dissolved gases in the process waters sometimes cause corrosion of the machinery, and carbon dioxide, in concentrations above 25 mg/l, has been reported to affect adversely the formation of sheets on the paper machines. The most undesirable gas is hydrogen sulfide. Even when present in very low concentration (0.3 mg/l or below) it is corrosive to metals and markedly shorten the life of the wire on Fourdrinier or cylinder machines.

The presence of micro-organisms in the process waters, such ad bacteria, algae, and fungi, may lead to difficulties from the growths of slimes, slick spots on the paper, and the development of odors. Odors may be taken up by the paper and should be avoided, especially when making paper products for food packaging purpose.

Some of the specifications approved as tentative standards by the Water Committee of the Technical Association of the Pulp and Paper Industries are summarized as follows:

TABLE – 79.1

SUMMARY OF SPECIFICATIONS FOR CHEMICAL COMPOSITION OF PROCESS WATERS FOR MANUFACTURE OF VARIOUS PAPERS*

Substance (amount given in mg/l)	Ground wood Paper	Soda and Sulfate Pulps	Kraft paper		Fine Paper
			Bleached	Unbleached	
Turbidity (as SiO_2)	50 (see note)	25 (see note)	40	100	10
Color (in platinum units)	30	5	25	100	5
Total hardness (as $CaCO_3$)	200	100	100	200	100
Calcium hardness (as $CaCO_3$)	–	50	–	–	50
Magnesium Hardness (as $CaCO_3$)	–	50	–	–	–
Alkalinity to methyl orange (as $CaCO_3$)	150	75	75	150	75
Iron (as Fe)	0.3	0.1	0.2	1.0	0.1
Manganese (as Mn)	0.1	0.05	0.1	0.5	0.05
Silica (as soluble SiO_2)	50	20	50	100	20
Total dissolved solids	500	250	300	500	200
Free carbon dioxide (as CO_2)	10	10	10	10	10
Chlorides (as Cl)	75	75	200	200	–
Residual chlorine (as Cl_2), max	–	–	–	–	2.0

Note: Materials causing turbidity shall not be gritty.

* Reproduced with the kind permission of the Resources Agency of California, State Water Resources Control Board, California from their publication "WATER QUALITY CRITERIA", 1963, edited by J.E. Mckee and H.W. Wolf.

CHAPTER - 80

RAYON MANUFACTURE

Rayon is a man made fibre resembling silk. It is actually regenerated cellulose. There are two types of viscose rayon, 1) Textile viscose rayon 2) High tenacity viscose rayon. Textile viscose rayon is the mostly manufactured item and is used primarily in women's apparels, upholstery and blended with wool in carpets and rugs. The high tenacity viscose rayon is used in cords for tyres, hose and belting.

The basic raw material used for the manufacture of viscose rayon is wood and cotton linters. The important chemicals needed are caustic soda, sulfuric acid, carbon-disulfide and zinc sulfate.

The cellulosic raw materials contain both alfa cellulose and hemi cellulose. Alfa cellulose is the basic component required for rayon manufacture. It is separated from the hemi cellulose by steeping wood or cotton in a solution of caustic soda (17 to 20% w/v). The alfa cellulose is insoluble while the hemi cellulose is soluble. The soluble portion is removed and recycled while the insoluble alkali cellulose is allowed to ripen for about 24 to 48 hours, at 75°F (This ripening or aging is allowed to get a solution after xanthation of correct viscosity for spinning). The aged alkali cellulose is converted to cellulose xanthate by treatment with carbon-di-sulfide under carefully controlled temperature and reduced pressure for two hours with churning. The cellulose xanthate is dissolved in caustic soda solution (2 to 2.5% w/v) to yield a proper viscose solution. The resulting solution known as viscose is separated by filter pressing, deaerated at high vacuum and at a temperature below room temperature to remove small air bubbles (which may either weaken the final yarn or cause breaks during spinning) and sent for extrusion.

Viscose is extruded into fine jets under pressure through thimble like spinnerette containing 750 to 2000 holes into the spinning bath containing dilute sulfuric acid (8% w/v), zinc sulfate (0.5 to 17% w/v), glucose and other chemicals which are essential to the spinning process. Sulfuric acid neutralizes free NaOH and decomposes xanthate liberating H_2S, CS_2 and S. The presence of 4% glucose prevents crystallization of salts in filaments.

The extruded yarn is washed to remove spinning liquor until free of acid with excess water. The sulfur adhering to the yarn is removed then by treatment with a very dilute solution

(1%) of sodium sulfide and again washed with copious volume of water. Then the yarn is bleached with hypochlorite solution, washed with dilute acid solution and detergents, dried properly and conned.

In the manufacture of rayon water is used primarily in the initial chemical solutions and in the finishing operations. Rayon manufacture requires water of standard quality. The water used for the manufacture of rayon (from wood cellulose or cotton linters) should be colorless and clear and fulfil certain requirements. They are briefly discussed in this chapter.

The recommended range of dissolved solids concentration is 100-200 mg/l.

Hardness in water is highly detrimental in the production of rayon. Eskell Nordell[*] describes the importance of softened water as follows:

"In the manufacture of rayon, it has been found extremely important to eliminate even a few ppm of hardness, and water with a practically zero content of hardness is a prime requisite.... In the manufacture of rayon, low-ash alpha cellulose is best produced either from pulp or linters in zeolite-softened water. Requirements for completely softened water in the subsequent manufacturing operations are so stringent".

Excess alkalinity is undesirable for the manufacture of rayon. The recommended range of alkalinity is 50 - 75 mg/l (as $CaCO_3$).

Heavy metals are also undesirable in water for rayon manufacture. The upper limit recommended for heavy metals in water to be used in rayon industries is 0.01 mg/l. Aluminum compounds are extensively used in water treatment and they gain access in treated water. Their presence in water is highly undesirable especially for rayon manufacturing and processing. The recommended maximum permissible concentration for aluminum is 0.25 mg/l. (This unit is more stringent if the water is used for dyeing rayon. The residual alum content, should not exceed 0.4 mg/l (as alum) or (0.04 mg/l as aluminum for dyeing.) The hydroxides/soaps of aluminum may produce stains).

[*] Eskell Nordell, "WATER TREATEMENT FOR INDUSTRIAL AND OTHER USES", 1951, Reinhold Publishing Corporation, New York.

Silica is not a troublesome constituent for other textiles. However its presence beyond a certain limit is reported to be detrimental for rayon manufacture. The recommended maximum upper permissible level for silica in water for rayon pulp production is 25 mg/l. If the natural water contains excess silica, it can be removed by ion-exchange method using highly basic anion exchange resins or by reverse osmosis. If colloidal silica is present it can be removed by lime-soda-magnesia treatment.

It should be ensured that water used for rayon manufacture is non corrosive.

Besides chemical limits, it is also desirable that the water should be free from algae and mould spores. It should also be of satisfactory bacteriological quality (i.e) when tested by Standard Plate Count method (at 37°C for 48 hrs), the bacterial content shall not exceed 100 per ml.

The recommended and threshold limiting concentrations of certain constituents of water are tabulated below:

TABLE – 80.1
RANGE OF RECOMMENDED THRESHOLD AND LIMITING CONCENTRATIONS FOR THE MANUFACTURE OF VISCOSE RAYON*

S. No.	Constituent	Recommended Limiting concentrations for water for Viscose Rayon manufacture, in mg/l	
		Pulp Production	Manufacture
1	Color (Hazen units)	5	–
2	Turbidity (NTU)	5	0.3
3	Hardness (as $CaCO_3$), mg/l	8	5
4	Iron (as Fe), mg/l	0.05	0.0
5	Manganese (as Mn), mg/l	0.03	0.0
6	Iron and Manganese, mg/l	0.05	0.0
7	Total solids, mg/l	100	–
8	Alkalinity (as $CaCO_3$), mg/l	50 – 75	–
9	pH value	–	7.8 – 8.3
10	Aluminium oxide	< 8	
11	Silica	< 25	
12	Copper	< 5	

* Reproduced with the kind permission of the Resources Agency of California, State Water Resources Control Board, California from their publication "WATER QUALITY CRITERIA", 1963, edited by J.E. Mckee and H.W. Wolf.

CHAPTER - 81

SILK REELING AND PROCESSING

Silk is a fine, strong, lustrous natural fiber. It is extruded by the silkworm when it spins its cocoon. Silk is closely allied to cellulose and resemble wool in structure but contains no sulfur.

The quality of water is a prime concern not only for processing silk fabrics, but also for reeling silk. When the quality of water is not suitable for reeling, considerable wastage of valuable silk occurs. With the intention to provide guidelines in the selection of suitable water and conditioning of water for effective silk reeling, some of the characteristics of water are discussed in detail in this chapter.

To ascertain whether a particular water is suitable for silk reeling, the following parameters have to be determined:

1) *Limpidity*
2) *Iron*
3) *Manganese*
4) *Hardness*
5) *Soluble Alkalinity*
6) *Nitrate*
7) *Nitrite*

LIMPIDITY

It is a term used to indicate the clarity of water. The water should be colorless and clear. It should not contain any suspended impurities either in solid or colloidal form. Silk has the tendency to absorb suspended matter present in water. This absorption is superficial and it affects the appearance of the fiber, particularly lustre- the precious characteristics of silk. Colored organic matter in suspension is especially undesirable as it spoils color as well as lustre of the silk.

IRON

When silk is reeled with water containing iron, its lustre is lost and the natural color is altered because of the ochreous deposit formed by the iron on the surface of the silk.

MANGANESE

It is reported to discolor silk. The tolerable level of manganese in water used for silk reeling is 0.1 mg/l. If iron is also present, then the total amount of iron and manganese should not exceed 0.25 mg/l.

HARDNESS

The hardness of water affects only surface characters of raw silk such as color, lustre and not its dynamometric properties. Hardness is however reported to increase the breakage of silk in reeling and throwing operations. In tin weighting and dyeing, hardness causes uneven deposition of tin and hungry spots in dyeing. Silk consists of twin threads bound together by a gummy substance known as sericin (silk gum) which is soluble in water. This sericin must be completely removed from the silk for satisfactory dyeing. Reeling the silk in hot water removes most of the sericin. If reeling is carried out in hard water, calcium salts precipitate and adhere to the fibre leading to an unpleasant product and in uneven dyeing. Though boiling with soap solution results in the complete removal of sericin, the calcium soaps formed, penetrate in to the fibre spoiling the natural lustre, and dyeing may be defective.

Inspite of the demerits of the hard water described above, a certain amount of hardness (particularly of temporary hardness) is necessary in water for reeling. Moderately hard waters give better results than water of zero hardness, because of the following reasons.

(i) During treatment certain substances amounting to about 7% of the total weight are exuded by the cocoons, which impart acidity to water. This acidity is harmful to the color of the silk if it is naturally yellow. However this acidity is neutralized by the calcium salts present in water.

(ii) The solvent effect of hard water on the cortex is more than that of soft water.

To conclude, a water with a hardness of about 180 mg/l (as $CaCO_3$) of which two-thirds is of temporary nature* can be regarded as suitable for silk reeling.

* The hardness of water is of two types viz., temporary and permanent. Temporary hardness of water is due to the presence of calcium and magnesium bicarbonates which will be removed on boiling (or by the addition of lime water). Permanent hardness of water is caused by the sulfates and chlorides of calcium and magnesium. These can only be removed by passing through ion-exchange resins.

For the preparation of dye-baths, and in other processing, it is advisable to use soft water only.

SOLUBLE ALKALINITY

This is the alkalinity of water remaining after a prolonged boiling and is due to the presence of small quantities of calcium and magnesium carbonates which remain in solution and due to the presence of sodium and potassium carbonates.

In silk reeling this soluble alkalinity is an important criterion. This is expressed in terms of mg. Na_2CO_3 per litre. Water with a soluble alkalinity lower than 15 mg/l is regarded as having a lower soluble alkalinity; that between 15 to 40 mg/l as medium, and above 40 mg/l as strongly alkaline.

The 'Soluble alkalinity' should not be confused with 'alkalinity' generally used in water chemistry which is due to bicarbonate, carbonates and hydroxides of calcium ,magnesium, sodium and potassium and to certain extent due to borates, phosphates and silicates, whereas soluble alkalinity is only due to the carbonates of calcium, magnesium, sodium and potassium. This soluble alkalinity is determined by titrating 25 ml or 50 ml or suitable aliquot of water sample after boiling, with 0.02 N. H_2SO_4 using phenolphthalein and mixed indicators.

$$\text{Soluble Alkalinity (as } Na_2CO_3\text{) mg/l} = \frac{\text{ml } 0.02N.H_2SO_4}{\text{ml.sample taken for estimation}} \times 1.06 \times 1000$$

Note: Degumming of silk (scouring silk) is carried out by treating the silk for 2 hours (or more if necessary) at about 95^{0}C in a 0.5 to 0.75 percent solution of soap, using 30 litres of water per kg of silk. At this temperature most of the bicarbonates and carbonates of calcium and magnesium get precipitated. The remaining alkalinity at this temperature (95^{0}C) is the soluble alkalinity and hence this parameter has to be determined separately as described above.

Water having medium soluble alkalinity is desirable for silk reeling because of its solvent effect on cocoons.

Strongly alkaline water is detrimental for silk reeling for the following reasons:

(i) When the bave is spun from the cocoons, it unwinds with knots and ravels and thus produces shaggy silk.

(ii) Sericin is softened and the threads tend to stick together, leading to the defect known as appiccicat (plastering).

(iii) Apart from all, the yield of silk is stated to be reduced.

NITRATES AND NITRITES

Nitrates and nitrites are reported* to be very injurious in the dyeing of the silk fabrics.

The specific quality requirements of water for silk reeling and processing are tabulated. The water used should be conditioned to conform to these requirements and also to other requirements as described for cotton and synthetic textiles.

TABLE – 81.1
SPECIAL REQUIREMENTS FOR WATER USED IN SILK REELING & PROCESSING

S. No.	Characteristics	Acceptable limits
1	Color	Clear and bright
2	Suspended matter	Freedom from suspended matter is desirable. Even trace amount is detrimental.
3	Hardness (as $CaCO_3$), mg/l	180 (in which about 120 mg/l should be of temporary hardness)
4	Iron (as Fe), mg/l + Manganese (as Mn), mg/l	Not to exceed 0.25 mg/l
5	Soluble Alkalinity (as Na_2CO_3), mg/l	Desirable range 15 – 40
6	Nitrate	Should be absent
7	Nitrite	Should be absent

* R. Michael, "Water as the cause of Poor colors and Failures in Dyeing", Farb V. Chemischerin 89 Oct 1942; Water Pollution Abstracts 17 (May 1944).

CHAPTER - 82

SOAP MANUFACTURE

Soap is produced by reaction of caustic lye with vegetable oils. In the preparation of caustic lye, if hard waters are used, wastage of alkali occurs through precipitation of calcium and magnesium salts forming insoluble soaps that become included in the soap resulting in an inferior product. The same applies to the boiling of the soap if hard water be added at this stage. Boiling of the soap means the caustic lye and vegetable oils are boiled together to form soap. The boiling is essential in order to attain the proper degree of hydration and to neutralize the fatty acids of the oils. If any hard water is added at this stage, calcium and magnesium soaps formed which would be incorporated in the product soap. Hard waters therefore are highly undesirable for soap manufacture.

CHAPTER - 83

SOFT DRINKS MANUFACTURE

The water requirement in a soft drinks manufacturing unit falls under three categories.

1) *Water used in making the soft drinks.*
2) *Water used for bottle washing.*
3) *Water for boiler feed-water make up chilling plant and other general uses.*

Water quality plays a vital role in giving the required taste to the soft drink. The quality requirements for water used in a soft drink manufacture are more stringent than those for drinking water.

The water should be colorless, clear, and free from any odor, taste, iron and manganese and organic matter. It must be of satisfactory bacteriological quality. (complete absence of micro organisms). Fruit-juice beverages are particularly susceptible to micro organism growth*. Since alkalinity destroy the flavor by neutralizing the acidity of fruit extracts, uniformly low alkalinity (not more than 50 mg/l) in water is an important requirement. As regards hardness, soft water is preferable. (But in soda water making, hard water is the choice. A water containing calcium sulfate is believed to make the most palatable soda water). In a bottling plant, normally water is treated for complete sterility. In these instances water would contain a high chlorine residual. It is highly undesirable as it destroys some of the organic materials used for flavoring. Besides it would impart a chlorine odor and taste to the soft drink†.

The water used for bottle washing will be about 10 to 15 times as much water as required for making the soft drink. Water intended for bottle washing should also be free from color, turbidity odor, taste, organic matter, iron and manganese. Water of zero hardness is

* Normally the finished water is passed through activated carbon filters as a final precaution for the removal of chlorine and any residual taste and / or odors. Periodic sterilization of these activated carbon filters is essential to prevent the growth of microbiological organisms.

† In order to avoid the trouble due to chlorine, most of the bottling plants at present have ultraviolet sterilizers and ozonators.

desirable for washing to prevent 'cloudy appearance' and formation of spots in the bottle. Besides eliminating deposits, zero hardness water save detergents to a marked extent. Iron and manganese reduce the brightness of the bottle and bottle would be dull in appearance after washing.

In the pasteurizing operation the bottled product is moved through a chilling zone (to arrest the growth of spoilage organisms) and then to controlled temperature water bath where the soft drink is brought to about 70°C by keeping for a specified time, then moved again to chilling compartment and packed. The temperature in the pasteurizing section is maintained by circulating hot water. It is reported that zero- hardness water is essentially required for this operation to avoid spotting of the bottles or cans.

CHAPTER - 84

STARCH MANUFACTURE

In the process of manufacturing, starch is sensitive to coloration, and since the product must be perfectly white, the waters should be free from iron, manganese and organic matter. Iron and manganese are especially detrimental in that they give a yellowish discoloration to the starch.

In the production of starch, hard water increases the ash content. Therefore water intended for starch manufacture should be as soft and mineral-free as possible in as much as any precipitated salt will accumulate in the starch. It should also meet the requirements of being clear, colorless, odorless, tasteless and of approved bacteriological quality.

Sago an edible form of starch is manufactured and consumed as food. Since it is as a food item the quality of water used for the manufacture of sago has to meet the requirements of water used in a food industry. Refer Chapter – 68 “FOOD INDUSTRY” for more details.

CHAPTER - 85

STEEL INDUSTRY

Iron and steel industries require more water than any other manufacturing industries. Approximately 120 tonnes of water is required for every tonne of steel produced. About 60% of water is exclusively used for cooling, the remaining used as process water. The following is a brief summery of the processes carried out in an integrated steel plant. Along with are given water requirements.

BLAST FURNACE OPERATIONS

In iron making, blast furnace† is the heart, in which iron ore, coke and lime stone are charged. Hot blast is injected from the lower portion of the furnace through a circumferential openings called tuyeres (see next paragraph) at about 30 – 35 lb/in^2 and 985^{0}C. Coke reacts with Fe_2O_3 and Fe_3O_4 releasing the iron with the production of carbon monoxide and carbon dioxide gases. The iron goes to the bottom of the furnace from where it is tapped periodically along with the slag. To keep this large furnace working efficiently, the equipments connected with the furnace are cooled at various points using water. The cooling water requirements is estimated to be about 65 – 950 litres / second.

The valves controlling the flow of hot blast air from the stoves to the furnace are cooled to prevent jamming or failure. The tuyeres are copper-jacketed nozzles with cooling water in the jackets. Since heat exchange rates are so high in the tuyere area it is important that cooling system be protected from fouling or plugging to maintain adequate heat transfer.

The cooling water has to be treated suitably to protect all these systems from corrosion, scale formation, and fouling from silt or microbial growths. It is common in the cooling systems that either one of these or any combination may exist.

Hot air blown through the furnace leaves at the top with a high velocity which is then passed on to a dry dust collector where all the entrained solid materials (burden fires ore, coke and limestone) are removed. Then the hot air proceeds to wet scrubbing system in which about 99% solids are removed. The scrubbing water is cooled down to reduce the temperature after the water contacts the hot gas. Cooling towers are employed for the recirculation of water. Huge volumes of water are required for cooling and needless to say a part of water

† The blast furnace is a very tall structure, the height goes upto 100 ft and its dia varies from 16 – 28 ft.

evaporates in the cooling tower rendering the remaining portion gets concentrated. This should be taken into account while designing the cooling water systems. The blow down schedules have to be programmed to control the salinity of the recirculating water and adequate chemical treatment is necessary to control scale formation and corrosion.

STEEL PRODUCTION

Iron emerging from the blast furnace is converted to steel by either one of the three methods namely,

i) *Basic oxygen furnace process*
ii) *Open hearth process, and*
iii) *Electric arc process.*

The choice of method depends on the final product required.

In all these methods most of the impurities of iron are removed to form steel. (Carbon is the most unwanted impurity in iron while making steel. Carbon content of iron is reduced from 4% to 0.2% to form steel). Cooling water is the major requirement in all these processes and very large volumes are required. To conserve water, only closed recirculating cooling water systems are installed. Wide variations in the water quality would normally occur and proper cooling water program has to be planned.

CONTINUOUS CASTING

Continuous casting is the process of continuously pouring molten metal from a ladle into the complex casting equipment. This method eliminates several steps in conventional steel production and effects a significant reduction in the overall cost. In this process also, large volumes of water are required for cooling. Appropriate water treatment and distribution are essential.

Unlike other cooling waters, the water used for cooling copper water–cooled mold (the very crucial point in the continuous casting) requires the highest quality of water. The cooling systems incorporates a closed loop with a secondary open cooling loop. Very low hardness waters, high purity boiler feed water or condensate are used as makeup. The harness of water should never exceed 10 mg/l. Since the system is a closed one, there is only little loss of water and the best corrosion inhibitors and dispersants can be used.

Auxiliary mechanical equipments of the continuous casting machines are also cooled for which separate cooling water systems are installed. The water should meet the cooling water requirements.

ROLLING OPERATIONS

In rolling mills, steel is converted into the desired products. There are two types of rolling, namely, hot rolling and cold rolling. The hot mill produces sheets, plates, bars and rods. Rolling is the water intensive operation in a steel mill. The hot rolling requires about 3,00,000 litres of water per tonne of hot metal rolled. Almost all the water is used for cooling. Cooling water is used to cool the doors and frames of furnaces. Besides, a large quantity of water is used for descaling the metal when it leaves the furnace to remove the oxide scale. Descaling water at a pressure of about 2,000 lb/ in^2 is blasted onto the surface of the metal so that scales are removed and no imperfection is caused. Water is also used for roll cooling and for spraying directly onto the steel. This apart, a considerable quantity of water is also used to cool the electric motors.

Most of the cooling water is recycled in a steel mill. The treatment program should be carefully drawn to eliminate potential problems of scaling, corrosion, fouling and microbial activity. This is especially important for water used in high pressure sprays.

Cold rolling mills do not require much water* as hot rolling. They are classified into two categories namely, single stand and multistand. In both the processes, water along with lubricant is used for cooling and to obtain a good finish. Special attention to water quality especially with respect to clarity, total suspended solids, iron and oil is also called for while planning a treatment program.

OTHER PROCESSES

In addition to the processes described above, certain other operations are also carried out in a steel mill. They are: Heat treatment, Sintering and Acid pickling.

HEAT TREATMENT

For the production of certain special grade steels with special physical properties, the metal is passed through a series of heat treatment systems. The major use of water in heat treatment is quenching.

* The water requirement in a cold rolling mill is estimated to be about 750 litres – 6000 litres per tonne of steel processed.

SINTERING

Sintering is a process employed to recover solid residues from scrubbers and clarifiers. Solid residues are iron ore fines, mill scale, limestone and coke fines. Water is used in this process to provide a moisture content of 5 – 8%. It is also used for spraying for dust control and to scrub the sinter furnace exhaust gases.

ACID PICKLING

In acid pickling, steel is treated with acid in acid baths to remove oxide from the metal surface thereby to produce a bright steel. Acid pickling thus renders the steel suitable for further finishing operations such as plating, galvanizing or coating. For pickling, hydrochloric and sulfuric acids are used. The major use of water in this operation is for rinsing the metal to render it free from acids.

UTILITIES

The water requirement for utilities in a steel industry are similar to other industries. Water is utilized mainly for steam generation and cooling. High quality water is required for boiler makeup. Cooling water is used for condensing turbine exhaust, cooling compressors and bearings of heavy machineries, the quality of which has to meet general cooling water requirements. A considerable quantity of water is also used for conveying ash from coal-fired furnaces to ash dykes in a steel industry.

The quality criteria of water desirable for various operations in a steel industry are not easily definable. However, based on the various studies conducted and field observations made at USA, the characteristics[*] that determine the desirable quality of water for steel manufacture are listed below.

1. *Temperature preferably below 75^{0}F.*
2. *Chlorides preferably below 175 ppm.*
3. *pH between 6.8 and 7.0.*
4. *Harness below 50 ppm.*
5. *Suspended matter below 25 ppm.*
6. *Organic content as low as possible.*
7. *Corrosion potential at the lowest possible level.*

[*] Reproduced with the kind permission from the Resources Agency of California, State Water Resources Control Board, USA from their publication J.E. McKee and H.W. Wolf "Water Quality Criteria", 2nd Edition, 1963

CHAPTER - 86

SUGAR INDUSTRY

Sugar industry is one of the largest water consuming industry. Sugar (sucrose) is derived from two major crops, namely sugar cane and sugar beet.

CANE SUGAR INDUSTRY

In the cane sugar mill, cane is chopped and shredded (after washing if necessary). The juice is then squeezed by passing it through crushers containing a series of roller mills. The cane is then wetted with water and the remaining juice is also extracted. The fibrous residue left over is called as bagasse which is used as a boiler fuel or sent for manufacture of paper and boards. The extracted juice is screened to remove floating impurities and then milk of lime is added.

Through the major ingredient of the juice is sucrose other natural impurities are also present in it. In order to maximize the yield of sugar and to reduce the impurities producing color, odor and taste, lime is added which precipitates these impurities. Besides the pH of the juice is increased to 7.6 - 7.8 there by the inversion of sugar is prevented. Along with lime, other coagulant aids such as bentonite is also added to coagulate the colloidal impurities. The mixture is then heated (to about 102°C) using high pressure steam and allowed to settle. The clarified and bleached juice is sent for evaporation (bleaching is effected by treatment with sulfur dioxide). The residue is sent to filter press (or rotary drum vacuum filter) and the juice in it is reclaimed. The mud from the filter press known as 'press mud' or 'lime mud' is used as a manure. The filter press is then washed thoroughly.

The clarified juice contains about 85% water. It is evaporated to approximately 40% water in multiple effect evaporators to a thick, pale yellow juice. This syrup is then fed into a single effect evaporator where the sugar is crystallized. (The latter operation is termed as 'pan boiling'). The mixture of syrup and crystal, known as massecuite is passed into a crystallizer which separates the crystallized sugar from the mother liquor. The spent mother liquor drains away separately as molasses. The sugar from the centrifugal basket is dried in a drier, graded and bagged.

All process water in sugar manufacture should be as free as possible from dissolved solids and devoid of suspended solids. Hard water is especially deleterious for it may form precipitates that accumulate in the refined sugar. Process water should be sterile not only with respect to coliform bacteria and pathogenic organisms, but also to saprophytic organisms that may cause destruction of sugar.

Recommended threshold concentration of various constituents in water to be used for sugar manufacturing are given below.

TABLE – 86.1
RECOMMENDED PERMISSIBLE CONCENTRATION OF CONSTITUENTS IN WATER USED IN CANE SUGAR INDUSTRY*

Constituent	Recommended Threshold values, in mg/l
Calcium	20
Magnesium	10
Sulfate	20
Chloride	20
Bicarbonate (as $CaCO_3$)	100
Iron	0.1

A special type of problem is associated with the condensate water in sugar industry. The condensate is normally used as boiler feed water. It is likely to contain ammonia in high concentrations and sometimes sucrose or invert sugar. The introduction of sugar into the boiler causes serious foaming problems and also quickly produces an acid condition which is detrimental to boiler system. Ammonia if present will corrode system that contain auxiliary equipment fabricated of copper alloys. Hence careful monitoring of the condensate for sugar content and ammonia is very essential to protect the boiler system.

Another problem associated with sugar industry is microbial infection which leads to serious problems. Sucrose is readily degraded by bacterial action to invert sugars, namely glucose and fructose. Under the conditions prevailing, they are further degraded to lactic acid in cane sugar industry and dextran in beet sugar industry. Since this degradation is primarily caused by bacteria, it is important to maintain control of microbial organisms through out the mill. Microbial activity is also reported to cause processing difficulties such as filter binding, production of odors and slime formation.

* Reproduced with the kind permission of the Resources Agency of California, State Water Resources Control Board, California from their publication "WATER QUALITY CRITERIA", 1963, edited by J.E. Mckee and H.W. Wolf.

Another important criterion is inversion of sugar by electrolytes. It is important that sugar refineries located near the sea coast should take precautionary steps to avoid sea water contamination.

BEET SUGAR INDUSTRY

Beer sugar industry is a heavy consumer of water. Water is not only used for extraction of sugar but is also used for cleaning the beets and for conveyance of the beets from the storage sheds to the factory and to carry away the spent pulp from the factory. Besides, large quantities of water are used for the production of steam. For sugar production, first the beets are sliced in to long narrow pieces of about one quarter inch thick. These are called "cosettes". The cosettes are dropped into a specially designed continuous counter current diffuser, where the sugar is extracted by hot water (70°C – 80°C). The resulting raw juice is blue-black in color with the sucrose content of 10 to 12%. (with a small amount of invert sugar and 2 to 3 % ash). The juice is then given a rough screening to remove coarser impurities. Next, milk of lime is added which precipitates most of the impurities. It is then filtered on pressure filters. The resulting filtrate is bleached with sulfur dioxide. The purified juice containing 10 to 12% sugar is concentrated to about 60% sugar in multiple effect evaporators. Then the procedure is similar to that of extraction of sugar from sugarcane.

Water quality plays a vital role in the production of beet sugar. The purity of sugar depends on the salts present in the water. The bicarbonates of calcium and magnesium present in moderately hard waters do not greatly interfere. However water containing a high concentration of calcium and magnesium bicarbonates may adversely affect the extraction of sugar from cosettes, as calcium carbonate and magnesium carbonate tend to precipitate at the temperature of extraction. The presence of chlorides and sulfates of calcium, magnesium and alkali metals in water is highly detrimental as these salts are reported to retain a notable quantity of sugar in the mother liquor. One part of salt will hold in solution five parts of sugar. Chlorides of sodium and potassium in the particular easily combine with the sugar, forming deliquescent compounds.

Water containing sodium chloride impart salinity and thus deteriorates the quality of sugar. This apart, there is a lurking danger of production of hydrogen sulfide, when waters containing sulfates of calcium and magnesium along with organic matter are used. In these cases, the sulfates are very easily reduced to sulfides. Water containing organic matter in suspension may also produce hydrogen sulfide. The produced hydrogen sulfide will spoil the sugar juice.

CHAPTER - 87

SYNTHETIC RESINS (PLASTICS) & OTHER POLYMERS MANUFACTURE

There are a wide variety of plastics and resins and a broad classification of them is given below with typical example in the parenthesis.

1) *Cellulosics (viscose rayon)*
2) *Vinyls (PVC)*
3) *Styrenes (polystyrenes)*
4) *Polyolefins (polythene)*
5) *Acrylics (methyl and ethyl acrylate)*
6) *Polyesters and Alkyd resins*
7) *Amino resins (urea formaldehyde)*
8) *Phenolic resins (phenol formaldehyde)*
9) *Miscellaneous resins (polyurethanes, nylon, polycarbonates etc.,)*

These are discussed briefly in the following sections.

CELLULOSICS

Wood and other cellulosic materials are digested to pulp and purified. This purified pulp is the major raw material for the production of cellulosics. They range from regenerated cellulose to nitro cellulose. Manufacture of the popular cellulosic material 'viscose rayon' and quality requirements of water for its manufacture are dealt exhaustively in Chapter – 80 "RAYON MANUFACTURE".

VINYLS

These are the most versatile thermoplastic resins made from compounds having a vinyl (- $CH = CH_2$) group. The important members of this class are polyvinyl chloride, polyvinyl alcohol, polyvinyl acetate and polyvinyl acetals.

Production of monomer is the first stage in the manufacture of vinyl resins. Then they are polymerized to the required extent and are separated from the unreacted monomer.

Finally the polymer is washed, dewatered and dried. Most of the processes do not involve the use of water but a few operations need water the quality which resembles to that of distilled water. The quality requirements of water used in a PVC industry is given below.

1) *pH value* *7.9*
2) *Total hardness (as $CaCO_3$), mg/l, max* *1*
3) *Phenolphthalein alkalinity (as $CaCO_3$), mg/l, max* *2*
4) *Total alkalinity (as $CaCO_3$), mg/l, max* *10*
5) *Chloride (as Cl^-), mg/l, max* *5*
6) *Iron (as Fe), mg/l, max* *1*
7) *Electrical Conductivity (μmhos/cm), max* *15*

STYRENES

Styrene monomer is produced by the pyrolysis – dehydrogenation of ethyl benzene (which is synthesized from ethylene and benzene). Styrenes are produced by mass polymerization and suspension methods. In these processes water is used as a raw material as well as for washing.

POLYOLEFINS (POLYETHYLENES)

Polythene is the first and is still the largest in the production of polyolefin plastics. Among the total production of synthetic resins, polythenes represent about 30% as they have wide applications. Since this reaction is also polymerization, water requirement is comparatively less.

ACRYLICS

Acrylic emulsions are the well known surface coatings (paints) used in industries and houses. Acrylic resins are produced by polymerization of methyl and ethyl acrylates and methyl, ethyl and butyl methacrylates. Three processes, namely, bulk, solution and emulsion polymerization are employed to obtain different types of products. Water is chiefly used for washing in these processes.

POLYESTER AND ALKYD RESINS

Polyester resins are complex esters formed by the reaction between a difunctional alcohol and dibasic acid (or anhydride). Unsaturated polyesters (laminates and molding compounds) are produced when any of the reactants contain unsaturation. Alkyd resins are polyester resins produced in the presence of oils or fatty acids.

In the manufacture of these resins also, the major use of water is washing.

AMINO RESINS

Urea–formaldehyde and melamine–formaldehyde are the two important amino resins. They are produced by condensation polymerization (40% aqueous solution of formaldehyde). In this process, water is used for cooling to control the temperature as the reaction is exothermic.

PHENOLIC RESINS

Phenolic resins are prepared from almost any phenolic body and aldehyde. Phenol-formaldehyde resins constitute the greater proportion. Other phenolic resins include phenol furfural and resorcinol-formaldehyde. Manufacturing processes and generation of effluents are similar to that of amino resins. Here also water is mainly used for cooling.

MISCELLANEOUS RESINS

This category includes, polyurethanes, nylon, polycarbonates, acetal, silicone and coumarone – indene resins. The production of these resins is carried out in a cooled recirculated system. Needless to say that the chief use of water is for cooling.

From the foregoing discussion it is clear that there are many types of synthetic resins and many complicated manufacturing processes. The quantity and quality requirements are based on the manufacturing processes. In general, water of unusual quality is not required. Where it is required in small quantities being employed in the process itself, demineralized water serves the purpose. Apart from manufacturing processes, water is mainly used for boiler feed and cooling. They have to meet the boiler and cooling water quality requirements.

CHAPTER - 88

SYNTHETIC RUBBER INDUSTRY*

The manufacture of synthetic rubber involves the polymerization of raw materials. In the synthetic rubber industry, large quantities of cooling water and smaller quantities of process water are required. Non-corrosiveness and freedom from suspended matter and algal growths are among the desirable qualities of water reported from a synthetic rubber plant in Pennsylvania. Water having a hardness of more than 50 mg/l and an oxygen demand greater than 3 mg/l, measured by potassium permanganate test was found to cause trouble in another plant in West Virginia.

* Reproduced with the kind permission from the Resources Agency of California, State Water Resources Control Board, USA from their publication J.E. McKee and H.W. Wolf "Water Quality Criteria", 2nd Edition, 1963

CHAPTER - 89

TANNERY

Tanning is the process of conversion of animal skins and hides to non-putrescible and tough material known as leather. The tanning processes are broadly classified as

A) Beam House Processing, and
B) Tan Yard Processing

There are two types of tanning, namely 'Vegetable Tanning' and 'Chrome Tanning'. In vegetable tanning, vegetable tan liquor is used, while chrome tan liquor containing chromium salts is used in chrome tanning.

A) BEAM HOUSE RPOCESSING

The dry matter of the skin is almost protein, reported to be about 85% collagen. The skin, also contains lipids, albumin, carbohydrates and globulin in minor amounts. In beam house processes, all the impurities are removed and collagen of the skin is made ready to absorb the tannin or chromium. A brief account of the beam house processes are given below.

(i) **Soaking:** The salted hides and skins are first soaked in water in pits (containing wetting agents) for oneday. This removes the salts, dirt, blood, dung and non-fibrous proteins and restores the moisture lost during preservation and storage.

(ii) **Liming:** After soaking, the hides are washed with running water and again soaked in a pit containing lime and sodium sulfite (8:1). Liming swells the hides/skins and loosens the hair. Further, the hides and skins become attractive and more amenable for the removal of trace protein impurities.

(iii) **Deliming:** The lime absorbed over the skins and hides are removed in this process by treatment with either ammonium chloride or sulfate. After deliming, the stock is sent for tanning.

B) TAN YARD PROCESS

(1) VEGETABLE TANNING

This type of tanning produces heavy leathers which are plumper, more easily tooled and embossed. The following are the sequential operations in vegetable tanning.

(i) **Tanning:** The delimed skins and hides are soaked in pits containing vegetable tan liquor (extracts of bark, wood, nut etc., containing natural tannins and also pyrogallol or catechol base) for 10 to 15 days.

(ii) **Myrobing:** The tanned materials are then soaked in myrobalan liquor.

(iii) **Oiling:** After myrobing, oil is applied liberally on the surfaces with suitable vegetable oil which makes the leather soft, pliable and resistance to tearing. The process of oiling is also termed as 'stuffing'.

(iv) **Drying:** The leather is dried and dyed.

(2) CHROME TANNING

It is carried out to prepare light and more resistant leather. It involves more operations than vegetable tanning, but the duration is much less.

(i) **Bating:** Bating is the treatment given to the delimed materials with a bating agent (mixture of prepared proteolytic enzymes) which reduces the pH and swelling, peptizes the fibres and removes the protein degradation products. Also renders the grain silky & slippery and smoother and porous.

(ii) **Pickling:** It is done to prevent precipitation of chromium salts on the skin fibres and involves pickling in sulfuric acid and common salt for one day.

(iii) **Tanning:** After pickling, they are tanned by soaking in chrometan liquor containing chromium sulfate for 6 hours. After tanning, sodium carbonate is added to the drum to fix the chrome. Then they are taken out of the drum and kept for 24 hours for complete absorption and fixation of chromium.

(iv) **Dyeing:** The tanned leather is then washed, neutralized and dyed with different types of dyes.

(v) **Fat – Liquoring:** To the leather is then applied an emulsion of sulfonated oil to make the leather soft and pliable.

Water is used in almost all the operations described above excepting a few. The quantity of water used in a tannery is very large and the impurities present in water that are precipitated by tannin have an important bearing on the success and economy of the tanning process.

The process water should be low in carbonate hardness, free carbon dioxide, color and turbidity and should be free from iron and manganese. Waters containing much carbonate are objectionable even in the preliminary washing of hides as calcium - carbonate tends to precipitate in the skin; moreover the washing process is also retarded. If water used in liming the hides contains an appreciable quantity of calcium bicarbonate, it gets precipitated and deposits in the crevices of dermic tissue. This deposits interfere with the absorption of the tannin in the cells of the hide and retard the tanning process causing brown stains on the leather (due to oxidation of tannate of lime). In the tan-pits, the calcium carbonate is transformed into tannate of lime which oxidizes rapidly in the air and forms secondary products of tan-oxylic or tan-omelanic acids giving a reddish brown coloration to leather which decreases its market value.

Magnesium bicarbonate also causes such effects in the tanning pits forming readily oxidizable tannates. Such precipitates would result in the loss of tannin, and by forming a coating on the leached material, interfere with the extraction. In view of such effects, it is preferable to use a water containing a carbonate hardness below 50 mg/l (as $CaCO_3$) for liming, deliming and bating.

Water rich in sodium and potassium carbonates are also reported to precipitate lime in the hides during liming process. Hence waters containing a total alkalinity not more than 150mg/l (as $CaCO_3$) is best suited for tanning purposes (especially for liming).

For leaching tannin extract, dyeing with basic dyes, and fat liquoring and for vegetable tanning, water of zero hardness is the best choice. However in the absence of such waters, the total hardness should not exceed 30 mg/l. Hardness has no apparent effects in chrome tanning and does not waste dyestuffs in acid baths. Hardness up to 300 mg/l is permissible for chrome tanning.

Water containing sulfates of calcium and magnesium are not harmful. Infact, sometimes they are added to water when they are deficient. Though chlorides are not harmful, their concentration is limited to 500 mg/l for both vegetable and chrome tanning as high amounts of chlorides tend to repress swelling of hides in tan liquors.

For tanning purposes, the water must be quite free from iron as it affects tanning. It forms dark colored precipitates with tannin, ranging from black to dark blue and green depending on the origin of the tannin. Such precipitates cause reduced tanning efficiency and results in discolored leather. Water containing not more than 0.1 mg/l iron (as Fe) is preferable for tanning. Manganese also acts in the same way and its presence is also objectionable. Manganese alone (as Mn) or in combination with iron should not exceed the concentration of 0.1 mg/l.

The desirable pH range of water for tannery uses is 6.5 - 8.0.

Waters contaminated with organic matter and bacteria are especially harmful in the soaking of hides as putrefaction is likely to set in and destroy the tissue of the hide. Water used for tanning when tested should not show a standard plate count more than 100 / ml.

CHAPTER – 90

TEXTILE PROCESSING
(COTTON & SYNTHETIC TEXTILES)

In a textile industry, yarn is prepared, woven or knitted and processed. The basic raw materials are cotton, synthetic fibres and wool. The processing methods vary from fibre to fibre. A brief account of such processes is given below.

COTTON TEXTILES

In a cotton textile mill, raw cotton is carded, combed, spun, slashed (sizing) and woven or knitted into cloth. All these processes are dry processes except sizing. The cloth obtained is called as grey cloth and is subjected to various treatments such as desizing, scouring, bleaching, mercerizing, dyeing (or printing) and finishing. These are wet processes and require a significant volume of water.

Sizing (Slashing): It is the process by which warp thread is sized with starch to give necessary tensile strength and smoothness required for weaving.

Desizing: It is a process employed to remove the sizing materials present in the grey cloth to make it suitable for dyeing and further processing. Desizing is carried out using either dilute sulfuric acid (0.5%) or with enzymes. Requirement of water is based on the desizing agent used.

Kier Boiler (Scouring): To remove the natural impurities such as grease, wax, fats and other impurities, the desized cloth is subjected to kier boiling, (i.e.) they are boiled with the aid of steam in an alkaline solution containing caustic soda (1 to 3%) (major component), soda ash, sodium silicate and sodium peroxide with small amounts of detergents for several hours.

Bleaching: The clothes are bleached to remove natural coloring materials and to render the cloth white. Alkaline hypochlorite or chlorine is used for bleaching cotton textiles. Bleaching is usually followed by washing with fresh water and then by scouring (treatment with sodium bisulfite) to remove traces of alkali and chlorine.

Some finer fabrics are bleached using peroxide.

Fabrics made from synthetic fibres are bleached using mild bleaching agents such as sodium hypochlorite.

Marcerizing: Mercerization involves treatment of the clothes with cold caustic soda solution (usually 20% w/v) followed by thorough washing with water. Mercerization gives a luster and strength to the fabric and mercerized fabrics are easily dyed. (Most of the clothes are usually mercerized but rough varieties are not mercerized, and after bleaching they are sent for dyeing without mercerizing).

Dyeing: The materials may either be dyed or printed. Dyeing is carried out with varied types of dyes. The commonly employed dyes are: i) Direct dyes ii) Basic dyes iii) Sulfur dyes iv) Vat dyes v) Naphthol dyes and vi) Developing dyes. Large volumes of water are used for dyeing.

In some of the mills yarn is dyed after scouring and bleaching.

Printing: It is carried out normally using hand blocks or in roller machines. The dye is applied as a thick paste along with starch or printing gums and mordents for fixation of dye. The fabrics are then thoroughly washed to remove the unfixed dyes etc.,

Finishing: This is done after dyeing or printing to impart smoothness to fabric. Starches, dextrins, natural and synthetic waxes and resins are used for finishing. The quantum of water used in this process is much less.

SYNTHETIC TEXTIELS

The most prominent man made synthetic fibres are, rayon, acetate, nylon, acrylic and polyester. These fabrics require no processing for the removal of natural impurities as they are man made. However they do require some treatments such as removal of size, antistat and lubricating oils, used for weaving operations.

The wet processes are generally similar for all these fabrics and include 1) Scouring and rinsing, 2) Dyeing or bleaching and rinsing, and 3) Final finishing (water proofing, shrink proofing etc.,)

WOOLLEN TEXTILES

The wool scouring and processing are dealt exhaustively as a separate chapter (see chapter – 92)

Water is one of the raw materials vital for all textile wet processes. Like alkalis, acids, dyes, mordants and other auxiliary chemicals, water is also an important raw material. The quantum of water used in a processing unit is many folds greater than any other chemical used, and naturally any impairment in quality of the water will affect the quality of the fabric. Treatment and conditioning of the water prior to processing, are very important in textile processing. The constituents of water and their effects on various textiles, namely, cotton and synthetic, are discussed in detail in the following sections.

COLOR

Color is normally considered as an indication of the presence of either soluble matter or suspended matter that may affect fibre / yarn / fabric, especially during dyeing and subsequent operations. Highly colored water are objectionable as some colors are taken up by the fabric / yarn resulting in off-tints and stains. While specific tolerances have not been established for all materials, it is suggested to keep the color within 5 color Hazen units, especially for special materials and for whites and delicate shades. This limit may be extended up to 20 Hazen units if dark or medium dark dyeing is done.

TURBIDITY

Turbidity is caused by the finer particles present in suspension. Turbidity in a water is highly objectionable for textile processing as the turbidity causing particles would be filtered on, while processing, and produce stains and / or streaking over the surface of the cloth or yearns. It is especially undesirable when package dyeing of yarns and fabrics (beam dyeing) is carried out. Color and turbidity can also cause dyespots and off shades (or similar defects). If the solid matter is colored, it is highly detrimental. Hence the quantity of solid matter is suspension, must be kept very low. (See also "Suspended matter"). Tolerances for turbidity may vary somewhat for different processes and for different materials, but in general should not exceed 5 NTU, and lower tolerances may be required for special materials.

Generally, surface waters are colored and turbid. It is to be mentioned that waters having the same turbidity may possess different types of solids in different quantities and their effects may vary. A knowledge on the turbidity causing particles is highly desirable in selecting proper treatment and treatment chemicals and to eradicate the nuisances caused by these particles.

SUSPENDED SOLIDS

Suspended solids as the name implies is the matter in suspension and no suspended impurity should be allowed in the process water as it is considered the major cause of staining and streaking. Supplies of water containing suspended solids may deposit solid matter on yarn and / or fabric. If the deposit is dark, stains occur or the dye is absorbed over the stains and the spots become more dark. This is quite common with moorland waters containing peaty matter in suspension. The well known kier stains on cotton are colored by dark brown humus particles, and are very difficult to remove owing to their resistance to wetting.

The prescribed limit for suspended solids in water is 5 mg/l. However, even a lower limit is required for some purposes such as package bleaching and package dyeing, as the packages form very efficient filters, and any deposition is shown as dark stains on the surfaces on standing or during drying of the bleached or dyed package. Hence to obtain water of utmost purity, the waters containing suspended impurities have to be subjected to coagulation and flocculation followed by pressure filtration.

[Most of the textile processing units in advanced countries carryout the following tests as a final check over the treated water. The final treated water is vacuum filtered through a fine paper membrane and the resulting stain is compared and measured against a white standard. If the intensity of the stain is more than a particular level, the water is again subjected to treatment]

It is to be mentioned that softened and clarified waters should be conveyed in pipes that will not corrode and add suspended matter to the water. PVC pipes are reported to be safer. Iron pipes though galvanized, are not preferred, as internal corrosion may take place in due course and rust produced thereupon would be transmitted to the process water.

DISSOLVED SOLIDS

The nature of soluble salts vary depending upon the source. Needless to say that waters containing excess dissolved mineral matter may deposit solids over the surface of the fabric. It is desirable that the dissolved solids content should be low and should not constitute a hazard. Various tolerance limits for dissolved solids are prescribed by different authors. J.M. Thompson in his book "Dye House Laboratory Practice", (1983), specifies a range of 65 – 150 mg/l as desirable. A.H. Little in the book "Water supplies and the Treatment and Disposal of Effluents" indicates that the waters containing dissolved solids below 500 mg/l are preferable for textile processing.

pH VALUE

Most of the textile processes are dependent on the pH value. A broad range of 7 – 9 for pH is prescribed for textile processing. However care should be exercised in keeping the optimum pH values for certain processes. During some of the processes, the pH may be shifted to above 9. For instance, at elevated temperatures, sodium bicarbonate present in the water is decomposed to sodium carbonate which increases the alkalinity. Such an increase in alkalinity may disturb some textile processes. Hence, along with the pH value, the quantum of alkalinity causing substances have to be considered together. If a water contains excess sodium bicarbonate, the upper limit for pH for such waters have to be kept below 9. In dyeing of nylon and acrilon, the pH of the water should be uniform because some of the dyes employed are pH sensitive*.

Most of the natural waters are slightly alkaline in nature. However, the upland waters and moorland waters are acidic in character and exhibit corrosive properties. These corrosive waters damage the fabric and have to be treated suitably. They are also undesirable as they affect pipes and other water supply structures, and in due course may carry over particles of rust and other corrosion products. A common type of corrosion that is frequently encountered in textile processing industry is the attack of iron in condensate return lines from dyeing machines. Under hot conditions, a rapid attack on iron steel and galvanized piping occurs, especially with waters of low pH (due to the presence of carbon dioxide), and the corrosion is accelerated by the dissolved oxygen present in the water. The corrosivity of water can be altered by the addition of alkaline materials such as lime, calcium carbonate of sodium silicate. Addition of lime or calcium carbonate may increase the hardness of water and selection of the material largely depends upon the quantity of water and the process for which water is required.

ALKALINITY

Alkalinity in water is not a specific substance but rather a combined effect of several substance and conditions. It is caused by the presence of bicarbonates, carbonates, hydroxides and to a certain extent by borates, silicates, phosphates and organic substances. Various limits for alkalinity in water have been prescribed y different authors, the maximum being 150 mg/l as $CaCO_3$, (as per the Bureau of Indian Standards). However the upper limit of alkalinity depends entirely on the intended use.

During cation exchange softening, a water acquires considerable amount of sodium bicarbonate** and sodium carbonate. At elevated temperatures, sodium bicarbonate

* W. Stump, Amer. Dyest. Rep 1966, 55, p 680

** The yellowing effect by sodium carbonate needs a special mention here. Sodium bicarbonate has an yellowing effect on clothes laundered in water containing it. This is ascribed to the formation of oxycellulose (in the presence of starch), when the bicarbonate concentrates under the heat of the iron. The decomposes to sodium carbonate, and carbon dioxide. The former increases the alkalinity further and may disturb some textile processes.

Softened water will generally have a high pH. For most of the textiles, it is harmless. However, for certain materials it is deleterious. Thus in the dyeing of rayon with certain direct colors (carried out in closed machines), the reducing substances present in the dye is dissolved by the alkalinity resulting in the reduction of some of the dye and destruction of color. (T.H. Mortan* hints out that this can be avoided by the addition of ammonium sulfate).

HARDNESS (CALCIUM AND MAGNESIUM)

Harness in a water is due to calcium and magnesium salts present in it. Eskel Nordell describes the detrimental effects of hardness in textile processing as follows:

"...... In cotton processing, hardness forms insoluble soap curds in kier boiling, soap scouring, soaping of fast colors etc; forms adherent deposits in mercerizing, preventing even penetration; forms deposits with alkaline "chemic"; and, especially in dyeing with substantive or direct colors, indanthrene dyes etc.,"

Calcium and magnesium salts will affect the efficiently of scouring and soaping-off process, and give rise to sticky deposits on the fabrics and also on machines due to the formation of calcium and magnesium soaps."

The absence of calcium is essential in water used for bleaching. Calcium causes soap wastage** and forms insoluble compounds with the bleach liquors which would adhere to the fibers. Formation of calcium soaps not only lead to the wastage of soap but also affect dyeing process. **Hard water is especially detrimental to vat dyeing operation and soft water is essentially required.**

The calcium soaps formed are insoluble and they produce a water repelling film over the fibre. The film is of a localized nature which cause resisting spots when the fabric is dyed.

---

effect is pronounced when the sodium bicarbonate alkalinity (as $CaCO_3$) exceeds 200 ppm; it is noticeable above 100 ppm but is unobjectiojable below 60 ppm.

* T.H. Mortan, "J.Soc. Dyers & Colorists', 1967, 83, 177

** It is estimated that for every 1000 litres of water and for each degree of hardness (17 ppm) the quantity of soap put out of action is about 100 gms. Despite the fact that synthetic detergents are not affected seriously to the extent of soaps, still the wastage is considerable.

The spot would appear as a white or lighter area after dyeing. In addition, calcium and magnesium ions precipitate certain dyestuffs leading to the wastage of dyes as well as causing uneven or streaky dyeing. Sometimes they acts as mordants, causing greater dyestuff absorption and deeper shades. Gallowing of fabrics is also attributed to hardness.

Calcium if present in excess quantities may be detrimental to a mordant bath, as if may precipitate and weaken the bath. It is also reported that it may change bichromate to chromate in a chrome mordant. (However, calcium is useful sometimes in the production of mordants for alizarin, haematoxylin and other types of dyeing with aluminum and iron salts. To produce necessary calcium concentration it is advisable to add calcium as calcium carbonate or calcium acetate to the water rather than to depend on the calcium originally present in the water).

Hardness is objectionable in linen processing too, especially in soda ash or ley boils, soaping, bleaching and many dyeing operations.

Despite the demerits, hardness has been described advantageous for certain processes. Some of the operations such as enzyme-desizing and peroxide bleaching and the rinsing of acid from fabric are reported to be more effective in moderately hard waters*.

METALS

This category includes iron, manganese, aluminium, copper and other heavy metals. They cause staining and / or interfere with the processes.

IRON AND MANGANESE (See Copper also)

Iron and manganese are extremely objectionable. Even the very low concentration of these metals stain the fabric. The hydroxides of iron, manganese (and aluminum) may produce stains either due to their color or capacity to bind colored solids. In some processes, iron and manganese exert catalytic effects. Either they will deposit on the textiles as hydroxides or combine with fatty acids to give metal soaps which cause staining and it is to be stated that these stains are very difficult to remove. With waters containing iron and manganese, dyeing becomes uneven and spotty. Various reports indicate that iron can seriously affect bleaching and dyeing operations.

* However, a high concentration of calcium and magnesium makes the peroxide bleach bath too stable. In most of the processing units, magnesium sulfate is added to the softened water to give the required degree of hardness and thus a separate supply of unsoftened water is eliminated.

Iron is reported to form lakes with many colors, particularly when a tannin mordant is used, the presence of iron is highly objectionable. In addition to spotting, iron may spoil the shade of the dyestuff, especially with mordanted goods. Even the minor concentration of iron (0.5 mg/l) and copper (0.1 mg/l) present in water can alter dye shades and cause cracking in latex backings on fabrics. When certain dyestuffs are used, the water should not be aerated for the removal of iron. Aeration may cause trouble by oxidation resulting in the wastage of dyestuff (particularly with indigo). The absence of iron is essential in water used for bleaching also.

While tolerance for iron are often stated as 0.2 to 0.3 ppm, even such lower concentrations are objectionable for many purposes. Manganese is even more objectionable as a concentration of 0.2 mg/l may cause dark brown or black stains on fabrics and tolerances usually should not exceed 0.05 mg/l.

When iron is present as ferrous bicarbonate, the water is generally aerated, (at pH 7 or above) allowed to settle and then filtered. Likewise manganese when present as manganese bicarbonate is usually removed by aeration (at pH above 9.0), followed by settlement and filtration. Organic iron and manganese are removed by aeration, coagulation with alum, settling and filtration.

ALUMINIUM

The aluminium content of natural water is negligible. However it enters the water supply on treatment with alum or other aluminum compounds. Waters containing aluminum are undesirable for textile processing as they form certain color lakes especially with special dyes such as alizarin. Though some authors prescribe a limit of 0.25 ppm for aluminium, it is desirable to keep the aluminium content within 0.1 ppm to protect from any undesirable effect. The residual alum (aluminium sulfate) is reported to affect the dyeing of rayon seriously. Hence for use in rayon, a residual content of aluminium not exceeding 0.04 mg/l as Al (or 0.4 mg/l as alum) is desirable.

COPPER

Copper is not a normal constituent of water. It is picked up during the passage of water through copper and brass fittings or pipes. The presence of copper in a water can seriously affect bleaching and dyeing operations. Hence it is advised to avoid copper / brass fittings especially in bleaching plants. Copper and iron are reported to cause instability in peroxide bleaching baths and damage the cotton during bleaching. Their deleterious effects can be reduced by the addition of sequestering agents such as the ethylene diaminetetra acetic acid or diethylene triamine penta-acetic acid to the bleach bath.

Copper is highly undesirable in water used for woollen industry. For details, refer Chapter – 92 "WOOL SCOURING AND PROCESSING".

SULFATE

Sulfates occur naturally in water, as a result of leachings from gypsum and other common minerals. They also enter in a water course through the discharge of industrial effluents. Sulfates also enter in a water supply on coagulation with alum. In moderate amounts, sulfates are not troublesome. However, for dyeing, it is desirable to keep the concentration reasonably constant so as not to affect the rate of dyeing. The recommended threshold concentration of sulfate in process water is 100 mg/l.

SULFITE AND SULFIDE

Natural waters do not contain any sulfite or sulfide. Paper mill effluents and dye house effluents contribute significant quantities of sulfite and / or sulfide to the receiving waters. They may cause difficulties in the processing, especially sulfides, which are reported to cause staining if present even in trace quantities. Sulfite and sulfide can be removed by acidification followed by aeration or by chlorination.

CHLORIDE

Chloride is the common anion found in all natural waters. They also enter in a water supply on contamination with sewage and industrial waste waters and through intrusion of sea water. Though chlorides do not cause any trouble in textile processing, they are undesirable as they appear to exert a significant effect on the rate of corrosion of steel and aluminium. Like sulfates, the main requirement with chlorides is to keep the concentration fairly constant, such that it will not affect dyeing. The upper permissible value for chlorides in process water is 100 mg/l.

NITRATE AND NITRITE

Small amounts of nitrates and nitrites are unlikely to cause any trouble in processing. However excess concentration of nitrate and nitrite can cause troubles in dyeing and can lead to staining with amino compounds. They are reported to be injurious to the wool in dyeing process.

ORGANIC GROWTHS

The presence of algae, moulds, slimes, iron bacteria etc., are very objectionable in process water. Organic matter and micro-organisms in the process water can be lead to the development of stains, odors and growths on the textile being treated. Hence the water should be free from all micro-organisms. The acceptable limits for the constituents of water intended for textile processing are tabulated below.

TABLE – 90.1
ACCEPTABLE LIMITS OF CHARACTERISTICS OF WATER USED IN TEXTILE PROCESSING

S. No.	Quality or Substance	Acceptable Limits (mg/l)
1	Color (Hazen)	Less than 10 units
2	Turbidity	Less than 5 NTU
3	Total Solids	Less than 500
4	Suspended Solids	Less than 5
5	pH Value	7 – 9
6	Acidity / Alkalinity	Less than 100 (as $CaCO_3$)
7	Hardness	Less than 70 (as $CaCO_3$)
8	Iron	Less than 0.3*
9	Manganese	Less than 0.05
10	Copper	Less than 0.01
11	Lead or heavy metals	Less than 0.01
12	Aluminium	Less than 0.25
13	Sulphate	100
14	Sulphide	Less than 1
15	Chloride	100
16	Nitrate	Less than 0.5
17	Chlorine	Less than 0.1
18	Ammonia	Less than 0.5
19	Oils, grease, fat, Soap	Less than 1

* A lower limit of 0.1 mg/l, for some uses.

Besides these limits, the water shall also be non-corrosive. It should also be free from algae, mould spots and organic growths.

CHAPTER - 91

THERMAL POWER PLANTS

(See also, Chapter - 75 "NUCLEAR POWER PLANTS)

The basic principle behind the generation of electricity is that when a magnet is rotated, an electric field is formed. So all the technologies in one or other way aim to rotate the magnet. This is accomplished by rotating a turbine to which the magnet is attached. Rotation of turbine is made by utilizing the force of water in a high place. Water from such place is brought down to a place of lower height through special cast iron pipes named 'penstocks' and used to rotate the turbines. The plants producing electricity in such a way are termed 'Hydro Electric Stations'.

Instead of channeling water through penstocks, water is converted to steam which rotates the turbine to generate electricity. The units producing electricity in this manner are called Thermal Power Plants. Water is converted to steam by the fossil fuels such as oil, coal and other petroleum products in these units. However water is also converted to steam by the heat generated by nuclear reactions. Such power generating plants are termed as 'Nuclear Power Plants'. This chapter deals with the quality requirements of water used in thermal power plants. Nuclear power plants are discussed as a separate chapter.

In both the thermal and nuclear power plants, high pressure and super critical boilers are employed to produce steam. Hence the quality requirements of water in the power industries are very stringent and dependent on the type of boiler employed. See Chapter – 47 "BOILER WATER REQUIREMENTS". While these boilers are operated, about 99% of the steam is returned as condensate and they require protection only from corrosion. However regular checking is also necessary to ensure against accumulation of dissolved solids as scales. This apart, enormous amount of water is also used for cooling (see Chapter – 48 "COOLING WATER REQUIREMENTS"). Some of the problems associated with power industry with reference to boiler water and cooling water and their treatment are discussed briefly in the following sections.

All the treatment programs in a power plant aim to prevent the tube metal failures in high pressure boilers. The degree of purity of water varies for different cycles in a power generation system. However one thing is certain, that relatively a high purity water is required.

The contaminants may be allowed only in ppb* levels when compared to ppm levels allowed in lower pressure cycles. Technology has given way to produce such high purity waters. Ion exchange and membrane processes are proved to be excellent methods. The major problems however is not the production of pure water but keeping the water pure after it has entered the cycle.

The purity of water is seriously affected by two major phenomena namely, corrosion and inleakage. Corrosion not only eats away the metal but the corrosion products enter into water as contaminant. These are subsequently deposited on heated boiler tube surfaces where they function as porous membranes. (See "ductile gouging" described in next paragraph). Inleakage-the second major source of contamination is a phenomenon by which low purity water enters the cycle at points of high vacuum (e.g) the surface condenser. Corrosion and inleakage however can be minimized by careful selection of the materials of construction (copper-nickel alloys and carbon steel is recommended).

In power plant drum type boilers, a highly localized water side corrosion namely, 'ductile gouging' occurs. The ductile gouging – a form of corrosion is a chemical attack characterized by irregular craters in the tube wall. The most common chemical found to cause such corrosion is sodium hydroxide. At vulnerable points (where the wall thickness is insufficient to withstand the internal pressures) a porous deposit may be formed on the tube of the wall. This deposit provides a mechanism for the concentration of sodium hydroxide to several thousand mg/l. Sodium hydroxide concentrates in this manner first attacks the tube surface magnetite (Fe_3O_4) and elemental iron of the tube to produce soluble iron complexes. Thus a significant wastage of metal occurs. During this reaction atomic hydrogen is produced which is capable of intergranular penetration of steel and reacts with the carbon of the steel to produce methane resulting in the decarburization and weakening of the steel. The produced methane is also harmful. As methane is produced continuously, a pressure is developed between grains in the steel microstructure resulting in the discontinuous intergranular cracks. The net result is the failure of tube walls. To prevent such corrosion and hydrogen damage, either all concentration mechanisms must be eliminated or potentially corrosive chemicals must be excluded from the boiler water.

* ppb – parts per billion (microgram / litre)

Feed water oxygen plays a vital role in pre-boiler corrosion and contamination. An oxygen level of 5 ppb or less is desirable. Oxygen is reduced to low levels mechanically using deaerators. The traces of oxygen remaining after deaeration is usually removed by the addition of organic reducing agent such as hydrazine*. Hydrazine leaves no solids to water.

Feed water pH is an important factor in minimizing pre-boiler corrosion of metals. A minimum pH of 8.5 is required for iron and its alloys whereas a pH value in the range of 9.2 – 9.6 is optimum for copper and its alloys. Where both metals are present, a balance has to be struck (i.e) a pH range of 8.8 – 9.2 is maintained. Ammonia and other volatile amines (which do not leave any solids) are generally employed to increase the pH value.

As discussed above, corrosion of boiler water is also a function of pH. The desirable pH range of the boiler water is 9.0 – 11.0. Such pH enhancement has to be achieved without the production of sodium hydroxide as described above. In utility boiler practice, this is accomplished by the addition of blends of sodium phosphate salts or by all - volatile treatment (AVT)*. The phosphate blends normally contains disodium and trisodium phosphates in the ratio of 2:3 to maintain the ration of Na : PO_4 not to exceed 2.6 to prevent the formation of sodium hydroxide and thereby to provide a positive protection against caustic – concentrate – type corrosion damage. Such addition of phosphates in coordination with pH is referred as 'coordinated phosphate – pH control of alkalinity'.

Ammonia is another important material that causes corrosion. It is generally present in power cycles either due to direct addition or decomposition of chemicals such as hydrazine or amines which are added to the system. If present in sufficient concentrations, ammonia reacts directly with copper to form soluble complexes. It is reported that even at lower concentrations, ammonia is able to react with protective copper oxides on the water side of condenser and heater tube surfaces. Such reaction would result in the exposure of the parent metal paving way for attack by oxygen. To reduce such effects, concentration of ammonia is limited to only about 0.1 – 0.3 mg/l by controlled chemical addition and by proper regeneration of condensate polishers. In order to avoid the possibility of any regenerant chemical entering into cycle water, the resins of condensate polishing system are removed and regenerated externally.

* Due to its carcinogenic properties, hydrazine is slowly replaced by patented products containing carbohydrazide and ascorbic acid.

* All volatile treatment uses only nitrogen – hydrogen compounds such as ammonia and hydrazine and no solids are added. This treatment is also called as 'zero solids treatment'.

It is reiterated again that condensate polishing is imperative in a power generation industry, as it eliminates most deposit forming materials that may cause chemical concentration mechanisms in steam generator tubes. However periodical cleaning for the removal of deposits (principally oxides of iron and copper) accumulated on boiler tube surfaces is also essential.

Hitherto, discussion on corrosion only has been made but that much importance is not given to scale formation. Needless to say that the makeup water is deionized thoroughly and made free of silica. Since the water is free from any measurable hardness and other solids, tube failure due to water side scale formation is less likely to occur in the boilers of power industry than in low pressure industrial boilers.

It should be mentioned that the cooling water requirements in power generation is exceedingly high when compared to any other industry. Once through cooling is not possible for all power stations excepting the ones located near sea as such great quantities cannot be tapped from any source excepting from sea. The cooling systems in power stations located near sea which utilize sea water are built in such a way to withstand corrosion. Other power plants located elsewhere mostly recirculate cooling water and cooling towers / spray ponds are essential, and these systems have to be protected from scale formation and corrosion and fouling by microbiological growths.

Another major use of water in coal –fired power plants (and to a lesser extent in oil-fired plants) is for ash transport. The solids that fall to the bottom of the boiler furnace (bottom ash)[1] are invariably cooled and conveyed from the boiler by water. The quantity of water used for ash handling is enormous. It is reported that about 10 mgd is required in a typical 1000 MW coal – fired power plants. As regards quality of water for these purposes, it is not an important criterion.

[1] There is another ash produced in coal-fired thermal power plants, popularly known as 'Fly ash' which is produced during combustion of coal and does not settle to the bottom because of its low weight and it removed either by electrostatic precipitators or bag filters or using wet scrubbers.

CHAPTER – 92

WOOL SCOURING AND PROCESSING

Wool is a natural (protein) fibre of animal origin obtained mainly from sheep's back. Of the wool removed from sheep, only 40% is usable wool fibre, and the remaining 60% is composed of impurities. Scouring of wool is carried out to remove all these impurities. After scouring, certain special processes are carried out to obtain a finished woolen cloth.

RAW WOOL SCOURING

Unlike cotton and synthetics (which are scoured only after weaving), the woolen fibre itself needs scouring and the scouring is termed as 'Heavy Scouring". In the scouring process, first the raw wool is steeped in a bowl containing water at a temperature of 32° to 42°C. The water soluble suint and other heavy dirt particles are removed at this stage. The water insoluble wool grease is removed by treating the desuinted wool with a mixture of detergent and sodium sulfate (or chloride) in two or three bowls (of capacity 5000 to 15000 litres) arranged in series. The operating temperature is 65°C. Practically all the natural impurities (mainly grease) and other impurities in wool are removed by this type of detergent scouring.

After scouring, the woolen fibres are rinsed in the bowls thoroughly with water at a temperature of 46°C to 50°C to remove the residual chemicals and remaining dirt.

WEAVING & FINISHING OPERATIONS

Stock Dyeing: Following scouring, dyeing is performed either in open or pressure type machines, after which the dyed fibre is rinsed thoroughly with water. The various classes of dyes used for woolen dyeing are direct dyes, acid chrome, pure mordant dyes and premetallised acid dyes. Besides, the indigo vat dyes, indigo and soluble vat dyes are also used under special circumstances.

Oiling: It is a process by which the cohesion of the fibres is increased. Oiling also aids in spinning. The oil used is olive oil or a mixture of lard and mineral oil. All the applied oil is washed out of the cloth during finishing.

After certain dry operations, the wool fibre is woven into fabric. The fabric undergoes certain wet processes to obtain the required finish and removal of traces of impurities. The sequence of operations are: fulling, carbonizing, piece dyeing (if stock dyeing is not carried out) and bleaching.

Fulling: This wet process imparts a felt quality to the wool. Further, the loosely woven wool is shrunk into a tight, closely woven cloth. Either acid fulling or alkali fulling is done. Acid fulling is normally carried out for heavy fabrics such as blankets, military uniforms, felt cloth etc., Alkali fulling is meant for other woolen fabrics.

In acid fulling, dilute solution of sulfuric acid and hydrogen peroxide are used along with small quantities of metallic catalysts. In alkali fulling, soap (or detergent) sodium carbonate and a sequestering agent are employed. After fulling, the fabric is washed extensively twice or thrice to remove the excess chemicals.

Wool Carbonizing: It is a final process applied for the removal of burrs and other impurities remaining in the wool. In this process, the woolen fabric is impregnated with sulfuric acid solution (4 to 7%) and then oven dried at 100° to $104^{\circ}C$ followed by mechanical agitation. The fabric is then rinsed with water, passed through baths containing sodium carbonate solution (2.5%) to neutralize the residual acid, washed again and then dried.

Piece Dyeing: The goods which have not been dyed at fibre stage are dyed after carbonizing.

Bleaching / Brightening: Sometimes, mild bleaching is also carried out with sulfur- dioxide or hydrogen peroxide to whiten the natural yellow tint of the wool.

The water used for wool scouring and processing should be colorless and clear. It should be non-corrosive. The recommended pH range is 6.0 – 8.5. The total alkalinity should be within 150 mg/l (as $CaCO_3$). Moderate hardness is desirable. Total absence of iron, manganese and copper is essential. The water should also be free from micro-organisms, algae and mould spores.

Some of the substances present in water that interfere seriously in wool scouring and processing are dealt elaborately in the following sections to provide more information.

HARDNESS

Hardness in a water is due to calcium and magnesium salts. The presence of hardness is undesirable in a water used for wool processing. Hard water forms insoluble calcium and magnesium soaps, which are readily retained in the wool surface due to the scaly structure of wool fibers. Hence all scouring and dyeing operations should be carried out only in zeolite – softened water.

Hardness in water, sometimes is advantageous in wool scouring. It is true that waters with zero hardness remove grease more readily from wool than a moderately hard water containing hardness 50 to 70 mg/l (as $CaCO_3$). However the moderately hard waters form lime soaps which possess the capacity to absorb dirt and fatty acids from the wool thus yielding a cleaner and softer product than with water of zero hardness. The optimum hardness required varies with the materials.

In wool bleaching with sodium peroxide, it is beneficial to add magnesium sulfate to water since it counteracts the effects of caustic soda produced from the peroxide.

COPPER

Total absence of copper is desirable. Copper is not a natural constituent of water. It is picked up by contact with copper and brass fittings or pipes. Contamination of copper more readily occurs with the use of ammonia solutions which attack copper / brass fittings rapidly on wool. For example, if the process water contains 5 mg/l copper (as Cu) the concentration on the treated fabric will be about 70 mg/kg. A published report* indicates that in the bleaching of wool, copper catalyzes the reaction with hydrogen peroxide (the bleaching agent) to the extent that holes are formed in the material. Hence it is imperative to take utmost care to avoid copper containing fittings in bleaching plant. Inspite of the precautionary measures copper enters the reaction, sequestering agents such as EDTA (Ethylene diamine tetra acetic acid) or Diethylene triamine penta acetic acids is added to bleach bath to prevent the activity of copper and iron. It is a usual practice to add calgon (sodium hexa meta phosphate) to wool scouring liquors to protect the fabric from the deleterious effects of copper and iron.

---

* Anon "Copper and Textile", Water Research News 4, 9 (1957) The Water Research Association, Red Hill, England

IRON (See 'Copper' also)

Iron also exerts rapid catalytic decomposition of the peroxide. Iron / Copper as metal or as their oxides or as their salts cause greatly accelerated local liberation of oxygen accompanied by serious tendering. They are reported to cause a number of pin holes in the fabric either visible immediately after bleaching or become apparent during wear. (The wool should also be free from particles of dirt containing iron because these become solubilized and spread making large brown stains). The maximum limit for iron is 0.1 mg/l. Since manganese also acts in the same way as iron, it should not exceed 0.1 mg/l alone or in combination with iron.

ALUMINIUM

Aluminium is also undesirable as it is likely to affect the wool during scouring and processing (especially in dyeing). The maximum permissible limit is 0.1 mg/l.

NITRATE AND NITRITE

Nitrates and Nitrites are reported to be injurious to the wool in dyeing process.*

MICRO-ORGANISMS

Wool is liable to be infected with micro-organisms especially during hot weather. Certain micro-organisms produce an enzyme which decomposes hydrogen peroxide (bleaching agent) causing very rapid evolution of oxygen and thereby resulting in ineffective bleaching. This trouble with micro-organisms can be eliminated by rescouring wool.

* R. Michael, "Water as the cause of Poor colors and Failures in Dyeing", Farb. U. Chemischrein, 89 (Oct 1942); Water Pollution Abs-17 (May 1944).

CHAPTER – 93

MISCELLANEOUS INDUSTRIAL USES

CITRUS FRUIT INDUSTRY

For citrus fruit industry, soft water is desirable. Soft water is more effective in cleaning the citrus fruits and improves the appearance of the product. In citrus industry, ion exchangers are employed for the treatment of peel juice to make pectin and citric acid.

CORN PRODUCTS MANUFACTURE

The water used in corn products manufacture should be clear, colorless, tasteless and of approved bacteriological quality. High magnesium content in the process water may lead to cloudiness in the corn syrup. The water may be softened (by ion exchange) to meet such requirements. The waters must also be free from iron and manganese.

DEXTROSE MANUFACTURE

In the manufacture of dextrose, demineralized water is reported to increase the yield and improve the quality.

GLUE MANUFACTURE

Glue is made from bone and odd ends of hide. Hard water retards the breaking up of the tissue. Therefore soft waters are preferable for the glue manufacture.

INDUSTRIAL COOLANTS
(WATER QUALITY FOR MIXING WITH SOLUBLE (MISCIBLE) OILS)

In industry, coolants are mainly used to cool the machinery parts and to reduce friction. A class of soluble (miscible) oils, are mixed with water and then used as coolant. In fact, the major part of the coolant is water and a fraction is the soluble oil. Only soft waters are used to makeup the coolants. If any hard water is used, calcium ions tend to destabilize the emulsions. Hence removal of calcium and magnesium in the makeup water extends the stability of the soluble oils. Besides quality of water to be mixed with oil to form coolant is very important as otherwise scales may deposit on the machinery parts or corrosion reactions may be induced.

The recommended values for certain parameters of water to be mixed with the soluble oils are given below.

Electrical Conductivity (μmhos/cm)	*: Below 1000*
Total Dissolved Solids (mg/l), max	*: 400*
pH value	*: Min 6.5 and Max 7.0*
Total Hardness, mg/l	*: Min 80 and Max 200*
Sulfate, mg/l	*: Below 50*
Chloride, mg/l	*: Below 100*

SILVER PLATING OF GLASS

Only distilled water has to be used for silver plating of glass for the manufacture of mirrors. Especially, the total absence of iron in waters used for the preparation of the silver solution is essential. The slightest trace of iron makes the silver deposit black and granular.

TELEVISION PICTURE TUBE MANUFACTURE

Manufacture of television picture tubes requires extremely pure water. Therefore ultra-pure water is the only choice. Television picture tube manufacturing units use ultra-pure water for washing, screening and lacquering processes. Any slightest impurity if present in these process waters would lead to the impairment of the quality of picture tube. The dissolved mineral matter, organic matter as well as iron and copper concentration of the water must be kept extremely low; especially iron and copper below 5 ppb. Even the slightest concentration of iron is reported to have an effect of deadening the phosphor. Copper is reported to cause a color shift. Other salts present if any would contaminate the phosphor. Some of them may form gases or produce air bubbles when the tube is hot ultimately resulting in tears or cracks in the phosphor screen. Therefore total elimination of electrolytes in the production water is extremely important.

YEAST MANUFACTURE

For the manufacture of yeast, distilled water is preferable. Sulfates of calcium and magnesium retard the fermentation. The sulfates and chlorides of sodium and potassium also retard fermentation to a lesser extent. Hence softening the water is not the method of treatment as it results in the alkali salts. Therefore it is economical to use distilled water for the manufacture of yeast.

CHAPTER – 94

ULTRA – PURE WATER

Ultra – pure water as the name implies is an extremely pure water essentially required for many industrial operations. The electrical conductivity of the water is 0.056 μmhos/cm (resistivity 18 megohm). The theoretical resistivity of pure water is 18 meghohm. Ultrapure water approaches this purity. Advancement in water purification technologies make it possible to produce a water of such high purity (99.999 percent purity) with a total impurity concentration down to 1 ppb*. Besides it is free from organic and particulate matter (which are not indicated by conductivity test) and micro-organisms and endotoxins.

The specifications for ultra-pure water have not been defined owing to the diverse applications of ultra–pure water in industries. For example, in the electronics industry even a trace of inorganic salt is not tolerable, while the pharmaceutical industry can tolerate chloride upto 1 ppm. An empirical criteria of ultra–pure water is given below. (Table 94–1). Also refer electronic grade water (Page No.125), water used in semiconductor processing (Page No.126) and water used in pharmaceutical industries (Page No.151).

TABLE - 94.1 EMPIRICAL CRITERIA FOR ULTRA - PURE WATER

S.No.	Parameter	Limit
1.	Electrical Conductivity	Not more than 0.056 μs/cm
2.	Silica (as SiO_2), ppb	Not more than 5
3.	Particle count, per ml	1
4.	Particle Size limit	Not more than 1 micron
5.	Total Organic Carbon	Shall not exceed 25 ppb
6.	Viable bacteria, maximum	1/1000 ml
7.	Endotoxins Eu/ml	Not more than 0.03

Ultra-pure water is used extensively in the electronics industry and in the preparation of pharmaceuticals. It is also used in nuclear power plants and as a makeup water for high

* ppb – parts per billion

pressure boilers. In the manufacture of television picture tubes also ultra-pure water is extensively used. The electronic industry, especially the manufacture of microchips needs the purest water of all.

In the electronics industry, ultra-pure water is used exclusively for cleaning the components such as cathode ray tubes, printed circuits, microchips, semi conductors, transistors and integrated circuits. The proper functioning and long term stability of electronic devices are reported to be enhanced by intensive cleaning with ultra pure water. It cleans the crystal surfaces absolutely such that the impurities (alkali and heavy metal ions) are removed completely. Ultra pure water is the only solvent that is able to provide such absolute cleaning. This apart, it is reported to possess several physical and chemical advantages.

The pharmaceutical industry is also a large user of ultra pure water. In the pharma industry, large quantities of water are used for the preparation, dissolution or dilution of medicaments for parenteral administrations. Besides, ultra pure water is also used for washing of surgical implant devices.

The use of boilers operating at or close to supercritical pressures (>3206 psi) is on the increase. They require water of very high purity (i.e.,) the impurity level should be in the range of ppb only. Therefore ultra–pure water is the choice for high pressure boilers. The efficiency of high pressure boilers is impaired by the contaminants of the feed water, if present even in very low concentrations. However, use of ultra-pure water with a total concentration of contaminants below 5 ppb eliminates such problems and it becomes possible to maintain the efficiency of high pressure boilers without any difficulty. This is especially true with the case of 'once–through boilers' (boilers without an effective drum).

Nuclear power plants require water of very high purity and ultra–pure water is the preferred one. It is required in 1) Water cooled reactors using water as a coolant and moderator or water–moderated reactors, 2) boiler operations 3) radiation shielding tanks and pools, and 4) various process operations.

Manufacture of television picture tubes requires water of very high purity. Even the slightest contamination would impair the quality and performance of picture tubes. Therefore in the manufacturing processes, namely washing, screening and lacquering, only ultra pure water is used. (For more details, see Page No.205)

The production methods of ultra-pure water are totally dependent on the end use of it. However the following is the general sequence of purification processes.

Raw water clarification → Depth filters → carbon filter → Reverse Osmosis plant → Deionizer (two or more stages of deionization) → Mixed bed Polishing → UV sterilization → submicron filter → Storage.

The order of location of RO plant and deionizer may vary depending upon the quality requirements. As regards electronics industry, the excess water used in the rinse operation is often quite pure. Therefore it is reclaimed, reprocessed and returned to the main storage tank. A special ultra–violet unit may be installed here for oxidation of organic matter at a wavelength of 185 nm.

Making ultra-pure water is relatively easy. However, the risk arises from recontamination. From the moment ultra-pure water is made, it is at risk from recontamination of every kind. Obviously the entire pure water circuit must be completely isolated from the surrounding environment. Besides, the materials of construction of the storage tank and distribution system are important criteria as they may contaminate water by leaching of organics from plastics or corrosion of metallic alloys. Some of the recommended materials for use in ultra-pure water system are listed below. 1) Teflon (tetrafluoro ethylene), 2) Tafzel (trifluoro mono-ethylene), 3) Tantalum, 4) Titanium 5) High purity zinc, 6) PVDF (Polyvinyledene fluoride), 7) Polyethylene, 9) Borosilicate glass 10) 304 L Stainless Steel 11) 304 stainless steel, 12) 316 L stainless steel, and 13) 316 stainless steel.

PART IV

DRINKING AND IRRIGATION WATER

CHAPTER - 95
DRINKING WATER

Water intended for drinking should be pure and wholesome. As defined in standard books, the 'Pure and wholesome waters' are the waters which are free from color, odor, taste, which contain no visible suspended matter, no excess of saline or organic matter, absence of bacteria especially coliforms. The waters must not have any appreciable action upon the materials of which mains, service pipes and fittings are made. However, the present day specifications are much more stringent owing to the high pollution levels observed in water courses due to the discharge of industrial effluents and sewage containing a number of organic and inorganic materials along with bacteria, viruses and other organisms and introduction of several pesticides for agricultural uses.

World Health Organization has taken into account all these facts and set limits for the substances that are considered toxic or that may cause ill health. Every substance that is likely to occur in water and its health impact has been studied in detail by WHO. Based on such studies, WHO has provided invaluable information on all the facts pertaining to drinking water by publishing three books, namely,

1) Guidelines for Drinking Water Quality, Vol – 1, Recommendations, Third Edition, 2004.
2) Guidelines for Drinking Water Quality, Vol – 2, Health Criteria and Supporting Information, Second Edition, 1996.
3) Guidelines for Drinking Water Quality, Vol – 3, Surveillance and Control of Community Supplies – Second Edition – 1997.

United States Public Health Service has also specified limits for many of the substances in drinking water. Based on WHO guidelines and USPHS standards, several countries have promulgated standards for drinking water quality.

While it is not possible to provide all the information pertaining to drinking water quality, certain microbiological aspects as described by WHO and limits prescribed by WHO for chemicals of health significance, for some of the organic constituents, for radioactive constituents, for substances and parameters that may give rise to consumer complaints are given below.

The tables provided here present a summary of guideline values for micro organisms and chemicals in drinking water. Individual values should not be used directly from the tables.

The guideline values must be used and interpreted in conjunction with the information contained in the text (WHO publications mentioned above).

MICROBIOLOGICAL ASPECTS*

Ideally, drinking-water should not contain any microorganisms known to be pathogenic. It should also be free from bacteria indicative of excremental pollution. To ensure that a supply of drinking-water satisfied these guidelines, it is important that samples should be examined regularly for indicators of faecal pollution. The primary bacterial indicator recommended for this purpose is the coliform group of organisms as a whole. Although as a group they are not exclusively of faecal origin, they are universally present in large numbers in the faece of man and other warm-blooded animals, and thus can be detected even after considerable dilution. The detection of faecal (thermo tolerant) coliform organisms, in particular Escherichia coli, provides definite evidence of faecal pollution.

Guideline values ensuring bacteriologically safe supplies of drinking water are provided in Table - 95.1.

It has been demonstrated that chlorination can produce virus-free water from faecally polluted source waters when the concentration of free residual chlorine is at least 0.5 mg/litre for a minimum contact period of 30 minutes at a pH below 8.0 and a turbidity of 1 nephelometric turbidity unit (NTU) or less. It is also desirable to maintain a free residual chlorine level of 0.2 – 0.5 mg/litre in the distribution system to reduce the risk of microbial regrowth. The detection of chlorine in this concentration range provides an indication of the absence of post-treatment contamination.

There should be an immediate increase in disinfectant application to achieve a free chlorine residual of 0.2 – 0.5 mg/litre in all portions of the distribution system when total coliform densities of more than 3 organisms per 100 ml are found in successive samples or when 1 or more faecal coliforms per 100 ml are detected.

Chlorine is pre-eminent as a disinfectant because of its ready availability and cheapness, as well as the ease with which it can be used, controlled and measured in water.

---

* Reproduced with the kind permission from World Health Organization, Geneva, from their publication "GUIDELINES FOR DRINKING WATER QUALITY", Vol – I, 2004 and Vol – III, 1997.

TABLE – 95.1

BACTERIOLOGICAL QUALITY OF DRINKING-WATER[a]

Organisms	Guideline value
All water intended for drinking	
E. coli or thermotolerant coliform bacteria[b, c]	Must not be detectable in any 100 ml sample
Treated water entering the distribution system	
E. coli or thermotolerant coliform bacteria[b]	Must not be detectable in any 100 ml sample
Total coliform bacteria	Must not be detectable in any 100 ml sample
Treated water in the distribution system	
E. coli or thermotolerant coliform bacteria[b]	Must not be detectable in any 100 ml sample
Total coliform bacteria	Must not be detectable in any 100 ml sample. In the case of large supplies, where sufficient samples are examined, must not be present in 95% of samples taken throughout any 12 month period.

a. Immediate investigative action must be taken if either *E. coli* or total coliform bacteria are detected. The minimum action in the case of total coliform bacteria is repeat sampling; if these bacteria are detected in the repeat sample, the cause must be determined by immediate further investigation.

b. Although *E. coli* is the more precise indicator of feacal pollution, the count of thermotolerant coliform bacteria is an acceptable alternative. If necessary, proper confirmatory tests must be carried out. Total coliform bacteria are not acceptable indicators of the sanitary quality of rural water supplies, particularly in tropical areas where many bacteria of no sanitary significance occur in almost all untreated supplies.

c. It is recognized that, in the great majority of rural water supplies in developing countries, feacal contamination is widespread. Under these conditions, the national surveillance agency should set medium-term targets for the progressive improvement of water supplies, as recommended in Vol – III "GUIDELNINES FOR DRINKING WATER QUALITY".

TABLE 95.2 CHEMICALS OF HEALTH SIGNIFICANCE IN DRINKING – WATER INORGANIC CONSTITUENTS*

	Guideline value (mg/litre)	Remarks
Antimony	0.005 (P)[a]	
Arsenic	0.01[b] (P)	For excess skin cancer risk of 6 x 10^{-4}
Barium	0.7	
Beryllium		NAD[c]
Boron	0.3	
Cadmium	0.003	
Chromium	0.05 (P)	
Copper	2 (P)	ATO[d]
Cyanide	0.07	
Fluoride	1.5	Climatic conditions, volume of water consumed, and intake from other sources should be considered when settling national standards.
Lead	0.01	It is recognized that not all water will meet the guideline value immediately; meanwhile, all other recommended measures to reduce the total exposure to lead should be implemented
Manganese	0.5 (P)	ATO
Mercury (total)	0.001	
Molybdenum	0.07	
Nickel	0.02	
Nitrate (as NO_3)	50	The sum of the ratio of the concentration of each to its respective guideline value should not exceed 1
Nitrite (as NO_2)	3 (P)	
Selenium	0.01	
Uranium		NAD

* Table 95-2 is reproduced with the kind permission from World Health Organization, Geneva from their publication, "Guidelines for Drinking Water Quality, Vol-2, Health Criteria and Supporting Information" – 2nd Edition - 1996

ORGANIC CONSTITUENTS

	Guideline value (μg/litre)	Remarks
Chlorinated alkanes		
Carbon tetrachloride	2	
Dichloromethane	20	
1, 1-dichloroethane		NAD
1, 2-dichloroethane	30[b]	For excess risk of 10^{-5}
1, 1, 1-trichloroethane	2000 (P)	
Chlorinated ethenes		
Vinyl chloride	5[b]	For excess risk of 10^{-5}
1, 1-dichloroethene	30	
1, 2-dichloroethene	50	
Trichloroethene	70 (P)	
Tetrachloroethene	40	
Aromatic hydrocarbons		
Benzene	10[b]	For excess risk of 10^{-5}
Toluene	700	ATO
Xylenes	500	ATO
Ethylbenzene	300	ATO
Styrene	20	ATO
Benzo[a]pyrene	0.7[b]	For excess risk of 10^{-5}
Chlorinated benzenes		
Monochlorobenzene	300	ATO
1, 2-dichlorobenzene	1000	ATO
1, 3-dichlorobenzene		NAD
1, 4-dichlorobenzene	300	ATO
Trichlorobenzenes (total)	20	ATO
Miscellaneous		
Di(2-ethylhexyl)adipate	80	
Di(2-ethylhexyl)phthalate	8	
Acrylamide	0.5[b]	For excess risk of 10^{-5}
Epichlorohydrin	0.4 (P)	
Hexachlorobutadiene	0.6	
Edetic acid (EDTA)	200 (P)	
Nitrilotriacetic acid	200	
Dialkyltins		NAD
Tributyltin oxide	2	

PESTICIDES

	Guideline value (µg/litre)	Remarks
Alachlor	20[b]	For excess risk of 10^{-5}
Aldicarb	10	
Aldrin/dieldrin	0.03	
Atrazine	2	
Bentazone	30	
Carbofuran	5	
Chlordane	0.2	
Chlorotoluron	30	
DDT	2	
1, 2-dibromo-3-chloropropane	1[b]	For excess risk of 10^{-5}
2, 4-D	30	
1, 2-dichloropropane	20 (P)	
1, 3-dichloropropane		NAD
1, 3-dichloropropene	20[b]	For excess risk of 10^{-5}
Ethylene dibromide		NAD
Heptachlor and heptachlor epoxide	0.03	
Hexachlorobenzene	1[b]	For excess risk of 10^{-5}
Isoproturon	9	
Lindane	2	
MCPA	2	
Methoxychlor	20	
Metolachlor	10	
Molinate	6	
Pendimethalin	20	
Pentachlorophenol	9 (P)	
Permethrin	20	
Propanil	20	
Pyridate	100	
Simazine	2	
Trifluralin	20	
Chlorophenoxy herbicides other than 2, 4-D and MCPA		
2, 4-DB	90	
Dichlorprop	100	
Fenoprop	9	
MCPB		NAD
Mecoprop	10	
2, 4, 5-T	9	

DISINFECTANTS AND DISINFECTANT BY-PRODUCTS

Disinfectants	Guideline value (mg/litre)	Remarks
Monochloramine di- and trichloramine	3	NAD
Chlorine	5	ATO. For effective disinfection there should be a residual concentration of free chlorine of ≥ 0.5 mg/litre after at least 30 minutes contact time at pH < 8.0
Chlorine dioxide		A guideline value has not been established because of the rapid breakdown of chlorine -dioxide and because the chlorite guideline value is adequately protective for potential toxicity from chlorine dioxide
Iodine		NAD
Disinfectants by-products	**Guideline value (µg/litre)**	**Remarks**
Bromate	25^{b} (P)	For 7 x 10^{-5} excess risk
Chlorate		NAD
Chlorite	200 (P)	
Chlorophenols		
2-chlorophenol		NAD
2, 4-dichlorophenol		NAD
2, 4, 6-trichlorophenol	200^{b}	For excess risk of 10^{-5}, ATO
Formaldehyde	900	
MX		NAD
Trihalomethanes		The sum of the ratio of the concentration of each to its respective guideline value should not exceed 1
Bromoform	100	
Dibromochloromethane	100	
Bromodichloromethane	60^{b}	For excess risk of 10^{-5}
Chloroform	200^{b}	For excess risk of 10^{-5}
Chlorinated acetic acids		
Monochloroacetic acid		NAD
Dichloroacetic acid	50 (P)	
Trichloroacetic acid	100 (P)	
Chloral hydrate (trichloroacetaldhyde)	10 (P)	
Chloroacetone		NAD
Halogenated acetonitriles		
Dichloroacetonitrile	90 (P)	
Dibromoacetonitrile	100 (P)	
Bromochloroacetonitrile		NAD
Trichloroacetonitrile	1 (P)	
Cyanogen chloride (as CN)	70	
Chloropicrin		NAD

[a] (P) – Provisional guideline value. This term is used for constituents for which there is some evidence of a potential hazard but where the available information on health effects is limited; or where an uncertainty factor greater than 1000 has been used in the derivation of the tolerable daily intake (TDI). Provisional guideline values are also recommended: (1) for substances for which the calculated guideline value would be below the practical quantification level, or below the level that can be achieved through practical treatment methods; or (2) where disinfection is likely to result in the guideline value being exceeded.

[b] For substances that are considered to be carcinogenic, the guideline value is the concentration in drinking–water associated with an excess lifetime cancer risk of 10^{-5} (one additional cancer per 100000 of the population ingesting drinking-water containing the substance at the guideline value for 70 years). Concentrations associated with estimated excess lifetime cancer risks of 10^{-4} and 10^{-6} can be calculated by multiplying and dividing, respectively, the guideline value by 10.

In cases in which the concentration associated with an excess lifetime cancer risk of 10^{-5} is not feasible as a result of inadequate analytical or treatment technology, a provisional guideline value is recommended at a practicable level and the estimated associated excess lifetime cancer risk presented.

It should be emphasized that the guideline values for carcinogenic substances have been computed from hypothetical mathematical models that cannot be verified experimentally and that the values should be interpreted differently from TDI–based values because of the lack of precision of the models. At best, these values must be regarded as rough estimates of cancer risk. However, the models used are conservative and probably err on the side of caution. Moderate short-term exposure to levels exceeding the guideline value for carcinogens does not significantly affect the risk.

[c] NAD – No adequate data to permit recommendation of a health-based guideline value.

[d] ATO – Concentrations of the substances at or below the health-based guideline value may affect the appearance, taste, or odour of the water.

TABLE - 95.3 RADIOACTIVE CONSTITUENTS OF DRINKING WATER[*]

	Screening value (Bq/litre)	Remarks
Gross alpha activity	0.1	If a screening value is exceeded, more detailed radionuclide analysis is necessary. Higher values do not necessarily imply that the water is unsuitable for human consumption
Gross beta activity	1	

[*] Table 95-2 is reproduced with the kind permission from World Health Organization, Geneva from their publication, "Guidelines for Drinking Water Quality, Vol-2, Health Criteria and Supporting Information" – 2nd Edition - 1996

TABLE – 95.4 SUBSTANCES AND PARAMETERS IN DRINKING-WATER THAT MAY GIVE RISE TO COMPLAINTS FROM CONSUMERS*

	Levels likely to give rise to consumer complaints[a]	Reasons for consumer complaints
Physical parameters		
Colour	15 TCU[b]	Appearance
Taste and odour	-	Should be acceptable
Temperature	-	Should be acceptable
Turbidity	5 TNU[c]	Appearance; for effective terminal disinfection, median turbidity ≤ 1 NTU, single sample ≤ 5 NTU
Inorganic constituents		
Aluminium	0.2 mg/l	Depositions, discoloration
Ammonia	1.5 mg/l	Odour and taste
Chloride	250 mg/l	Taste, corrosion
Copper	1 mg/l	Staining of laundry and sanitary ware (health-based provisional guideline value 2 mg/litre)
Hardness	-	High hardness: scale deposition, scum formation, low hardness: possible corrosion
Hydrogen sulfide	0.05 mg/l	Odour and taste
Iron	0.3 mg/l	Staining of laundry and sanitary ware
Manganese	0.1 mg/l	Staining of laundry and sanitary ware (health-based guideline value 0.5 mg/litre)
Dissolved oxygen	-	Indirect effects
pH	-	Low pH : corrosion High pH: taste, soapy feel preferably < 8.0 for effective disinfection with chlorine
Sodium	200 mg/l	Taste
Sulfate	250 mg/l	Taste, corrosion
Total dissolved solids	1000 mg/l	Taste
Zinc	3 mg/l	Appearance, taste

* Table 95-2 is reproduced with the kind permission from World Health Organization, Geneva from their publication, "Guidelines for Drinking Water Quality, Vol-2, Health Criteria and Supporting Information" – 2nd Edition - 1996

	Levels likely to give rise to consumer complaints[a]	Reasons for consumer complaints
Organic constituents		
Toluene	24-170 μg/l	Odour, taste (health-based guideline value 700 μg/l)
Xylene	20-1800 μg/l	Odour, taste (health based guideline value 500 μg/l)
Ethylbenzene	2-200 μg/l	Odour, taste (health-based guideline value 300 μg/l)
Styrene	4-2600 μg/l	Odour, taste (health-based guideline value 20 μg/l)
Monochlorobenzene	10-120 μg/l	Odour, taste (health-based guideline value 300 μg/l)
1, 2-dichlorobenzene	1-10 μg/l	Odour, taste (health-based guideline value 1000 μg/l)
1, 4-dicholorobenzene	0.3-30 μg/l	Odour, taste (health-based guideline value 300 μg/l)
Trichlorobenzenes (total	5-50 μg/l	Odour, taste (health-based guideline value 20 μg/l)
Synthetic detergents	-	Foaming, taste, odour
Disfectants and disinfectant by-products		
Chlorine	600-1000 μg/l	Taste and odour (health-based guideline value 5 μg/l)
Chlorophenols		
2-chlorophenol	0.1-10 μg/l	Taste, odour
2, 4-dichlorophenol	0.3-40 μg/l	Taste, odour
2, 4, 6-trichlorophenol	2-300 μg/l	Taste, odour (health-based guideline value 200 μg/l)

[a] The levels indicated are not precise numbers. Problems may occur at low or higher values according to local circumstances. A range of taste and odour threshold concentrations is given for organic constituents.

[b] TCU, true colour unit.

[c] NTU, nephelometric turbidity unit.

It has to be indicated that the water should be also of satisfactory biological quality. The drinking water shall be free from microscopic organisms such as algae, zooplanktons, flagellates, parasites and toxin-producing organisms.

CHAPTER - 96
IRRIGATION WATER

More fresh water is used for agriculture throughout the world – the largest portion being used for irrigation. Other agricultural uses include water supply for livestock, for the preparation of fertilizer and pesticide solutions and for diesel engine cooling. This chapter discusses exclusively the quality requirements of water for irrigation. The effects of substances present in water intended for irrigation may be discussed under the following heads.

1) *Salinity problems*
2) *Sodium hazard*
3) *Bicarbonate effects*
4) *Effects of Boron.*

SALINITY PROBLEMS

Owing to the many variable factors, no rigid limits of salinity can be set for irrigation waters. According to the authorities in Western Australia, water containing upto 1000 mg/l of salinity (dissolved solids) is suitable for growing all types of plants including the salt susceptible ones, provided the drainage is good. Water containing upto 2150mg/l is suitable for most plants except sensitive ones. Water containing upto 3150mg/l has been used for growing tomatoes, cabbages and other salt resistant plants. Generally, 3150 mg/l is near the maximum for the safe watering of any plant, and in such instances drainage must be excellent. Good drainage of the soil may be a more important factor for crop growth than the salts in the irrigation supply. Even when excellent waters are used, poorly drained land may sometimes go out of production.

Moderate concentrations of chloride in the root zone (700 to 1500 mg/l in the soil moisture) usually cause chloride to accumulate in the leaves to about 1 to 2 percent of dry weight. At such concentrations, marginal leaf burn develops, leading ultimately to leaf drop, twig die-back and possibly death of the plant.

SODIUM HAZARD

Calcium and magnesium in the proper proportions maintain soil in good condition of tilth and structure, while the opposite is true when sodium predominates. In most normal soils (in arid and semi arid regions), calcium and magnesium are the principal cations held by the soil in replaceable or exchangeable forms. When water containing high amounts of sodium is used for irrigation, it replaces calcium and magnesium from the soil, resulting in the break down of granular structure of soil. This apart, various changes takes place resulting in the scaling of soil pores and a decrease in soil permeability. With further increase in the sodium percentage, the soil continues to deteriorate and its pH increases to the level of alkali soils.

The percentage of sodium in water is calculated from the following formula.

$$Percent\ sodium\ (asNa) = \frac{Na \times 100}{Na + Ca + Mg + K}$$

The concentration of ions are expressed in milliequivalents per litre.

It has been widely recommended that the percent sodium in irrigation water should not exceed 50 - 60, in order to avoid the deleterious effects on soil as mentioned above.

Another type of sodium hazard is expressed in terms of Sodium Adsorption Ratio (SAR).

$$Sodium\ Adsorption\ Ratio\ (SAR) = \frac{Na}{\sqrt{\frac{Ca + Mg}{2}}}$$

where the ionic concentrations are expressed in terms of milliequivalents per litre.

A high SAR value affects the plant itself. Besides, it affects clay minerals with ion exchange capacities (causing the trace metals essential for crops to be regenerated off the clay particle). The permeability of clay soils is also reported to be reduced. Another report indicates that waters with a SAR value greater than 10 will present an appreciable sodium hazard in fine-textured soil having high cation-exchange capacity, especially as the salinity increases. However, there is no definite limit for SAR. It depends on the crops cultivated For example, the SAR limits for certain fruits may be 4 whereas grain and alfalfa may tolerate an SAR as high as 18.

BICARBONATE EFFECTS

The sodium hazard is also increased if the water contains a high concentration of bicarbonate ions, for as the soil solution becomes more concentrated there is a tendency for calcium and magnesium to precipitate as carbonates and for the relative proportion of sodium to be increased as a consequence. The bicarbonate values are conveniently expressed in terms of 'Residual Sodium Carbonate' (RSC).

$$RSC = \left(CO_3^{2-} + HCO_3^{-}\right) - \left(Ca^{2+} + Mg^{2+}\right)$$

where the ionic concentrations are expressed in terms of milliequivalents per litre.

Waters containing less than 1.25 meq/l of residual sodium carbonate are probably safe; those containing 1.25 - 2.5 meq/l are marginal; and those with more than 2.5 meq/l are not suitable.

EFFECTS OF BORON

The occurrence of boron in natural waters is rare. Although traces of boron are essential for all plant growth, it is doubtful whether more than 0.5 mg/l can be applied continuously to soils without ultimately producing some plant injury. Symptoms of boron injury can be distinguished easily from other types of injury. Among trees, advanced damage will result in leaf-yellowing and burning, pre mature leaf drop and reduced yield.

The quality requirements of water used for irrigation have been dealt exhaustively by Wilcox. Permissible limits for certain parameters quoted by Wilcox* and schofield† are shown in Table – 96.1

---

* Wilcox, T.a.Agricultural uses of reclarimed sewage effluents, "Sewage Works Journal", 20 (1948) 24 – 35.

† Schofield, C.S., Salinity of Irrigation Water, "Rep.Smithson. Instn: (for 1935), 1936, 275 – 87.

TABLE – 96.1*

PERMISSIBLE LIMITS FOR SOLUBLE SOLIDS, SODIUM, AND BORON IN IRRIGATION WATER

Class of Water	Soluble Solids, ppm	Sodium percent	Boron, ppm		
			Sensitive Crops[1]	Semi – Tolerent Crops[2]	Tolerant Crops[3]
Excellent	< 175	<20	< 0.33	< 0.67	< 1.0
Good	175 – 525	20 – 40	0.33 – 0.67	0.67 – 1.33	1.0 – 2.0
Permissible	525 – 1400	40 – 60	0.67 – 1.00	1.33 – 2.00	2.0 – 3.0
Doubtful	1400 – 2100	60 – 80	1.00 – 1.25	2.00 – 2.50	3.0 – 3.75
Unsuitable	> 2100	> 80	> 1.25	> 2.50	> 3.75

[1] Examples : fruit trees and bushes

[2] Examples: tomato, wheat, barley, olive, pea, potato

[3] Examples: carrot, lettuce, cabbage, turnip, onion, sugarbeet

Another classification based on percent sodium, boron, chloride, sulfates, electrical conductivity and total salts is given in Table – 96.2

TABLE – 96.2**

SUMMARY OF CLASSIFICATION OF IRRIGATION WATERS

Class	Na% $\frac{Na \times 100}{Na+Ca+Mg+K}$	Boron, mg/l	Chlorides in meq/l	Sulfates in meq/l	Electrical Conductivity in μ mho/cm	Total salts in mg/l
I	Less than 30 – 60%	Boron concentration recommended of this class of water is 0.5 mg/l; however tolerant plants will not be injured by 1 – 1.15 mg/l	Less than 2 – 5.5	Less than 4 – 10	Recent papers suggested a limit of 1000	Up to about 700
II	30 – 75%	0.5 – 2.0 mg/l, although for tolerant plants water with boron upto 3.35 mg/l may be satisfactory.	2 – 16	4 – 20	500 - 3000	350 - 2100
III	More than 70 – 75%	More than 2 mg/l although water with more than 1.0 may be highly unsuitable for sensitive plants.	More than 6 – 16	More than 12 – 20	More than 2500 - 3500	More than 1750 - 2100

* Table 96 – 1 is reproduced from "River Pollution II. Causes and Effects" – 1962, authored by LouisKlein. (Courtesy - M/s. Butterworth & Co., (Publishers) Ltd., London,)

1, 2, 3 – Eaton F.M., Boron in Soils and Irrigation Waters and Its Effects on Plants, "Tech. Bull. U.S.Dept., Agric. (1935) 448

** Table 96-2 is reproduced with the kind permission from the Resources Agency of California, State Water Resources Control Board, California from their publication "WATER QUALITY CRITERIA", 1963, edited by J.E. Mckee and H.W. Wolf.

PART V

IMPURITIES PRESENT IN WATER AND THEIR EFFECTS ON WATER TREATMENT SYSTEMS

CHAPTER - 97

IMPURITIES IN WATER AND THEIR EFFECTS ON WATER TREATMENT SYSTEMS

The impurities present in water and their effects on industrial uses have been discussed hitherto. The impurities not only affect individual processes but they do affect the water treatment systems too. Ion exchangers and reverse osmosis systems are severely affected. They are discussed in detail in the following sections.

ION EXCHANGE SYSTEMS

Natural waters and treated waters contain many impurities that foul the ion exchange resins. Resins have to be protected from contamination, for efficient operation and longer life expectancy. Turbidity, iron, manganese, oils, organic matter and microbiological growth are the principal foulants. Free chlorine also affects ion- exchange resins.

TURBIDITY

Turbidity in water is highly undesirable as it tends to coat the surface of the resins, resulting in the loss of exchange capacity. Hence influent water should be free from turbidity.

IRON AND MANGANESE

Iron and manganese in excess of 0.1 mg/l causes fouling of cation resins*. When the iron and manganese content exceeds 0.5 mg/l, the ion exchange resin will be coated and the material will lose its effectiveness. Further, ferrous iron is adsorbed over the surface of the ion exchange resin and during regeneration it will not be removed completely; it tends to deposit as insoluble hydroxide within the ion exchange resin if the pH is shifted, and the precipitate will be mechanically locked within the structure. The net result is the loss of exchange capacity of the resin. Iron and manganese from water can be removed by passing water through a filter

* Anion resins can also be fouled by iron due to the presence of iron in regenerant solution.

containing catalyzed manganese dioxide granules which precipitate iron as ferrous hydroxide which subsequently is filtered out.

OILS

Presence of oil is also reported to foul the resins. Oil fouled resins can be cleaned by washing with caustic soda solution or with suitable surface active agents.

ORGANICS

Organic matter in water causes permanent fouling of anion exchange resins. The term 'organics' include a wide variety of organic matter occurring in water in nature. They include humic acid, fulvic acid, tannin, lignin etc. These acids get adsorbed on anion exchange resins but a portion of them do not get eluted during regeneration. Resins fouled with organic matter results in lower operating capacity and high rinse requirements.

Organic fouling of the resins may be detected by one or more of the following symptoms.

i) *More rinse time is required than the normal.*
ii) *Conductivity of output water may not fall below the stipulated figure.*
iii) *Residual silica content of the demineralized water increases.*

Since most organic matter is colloidal in nature, they can be removed during pre-treatment stage by coagulation and filtration. Use of activated carbon is also recommended to reduce the organic matter content in water. The best treatment for the removal of organics is passing influent water through the macroporous resins prior to softening / demineralization. (Reverse Osmosis is also an effective method for the removal of organics).

It is recommended to clean the ion exchange resins atleast once in three months to remove the organic matter. Brine cleaning and sodium hypochlorite cleaning are effective methods. In brine cleaning, the cleaning agent is caustic brine. It is a solution containing 100 g. sodium chloride and 20 g. caustic soda per litre. For effective cleaning a warm caustic brine solution (40°C) may be used. The process must be followed by a thorough rinsing of the resins. In sodium hypochlorite cleaning, first the resin is treated with one volume brine solution

(10%) and then treated with three bed volumes of dilute sodium hypochlorite solution (containing 1% available chlorine). Needless to say that the resin has to be rinsed thoroughly.

BIOLOGICAL GROWTHS

Bacteria and fungi are the principal biological contaminants that foul resins. The bead structure and porous nature of ion exchange resins provide an atmosphere conducive to the growth of these organisms. If they are allowed to grow, they block the pores of the resin. This apart, because of their capacity to trap gas bubbles, the net density of resins will be reduced resulting in resin bead floatation. The best way to eliminate these organisms in the feed water is ozonation, UV irradiation or reverse osmosis. Chlorine may also be employed, however the residual chlorine has to be removed before passing the water onto the resin. If bacterial slimes have developed within a softener, they can be eliminated by the application of liquid bleach along with regenerant each time the unit is regenerated. The coatings of bacterial slimes and algal slimes can effectively be removed by applying warm caustic soda solution to the resin and keeping the solution for several hours in the softener. Then the softener has to be backwashed thoroughly to remove the dead slime and alkali from the bed.

FREE CHLORINE

The most common cause of resin damage is the presence of chlorine in water. Chlorine levels below 0.2 mg/l or tolerable. Above 0.2 mg/l, chlorine attacks the cross- linking of the resin matrix resulting in physical degradation. This reduces the efficacy as well as the life span of the resin. The residual chlorine in water has to be removed by passing it through activated carbon filter.

POLYELECTROLYTES

These are high molecular weight organic substances added in small quantities to aid coagulation and settling in the pretreatment. Higher concentration of these electrolytes not only reverse the coagulation process but also cause irreversible fouling of both cation and anion exchange resins. The nature and degree of fouling depend upon the polyelectrolyte used and its concentration.

REVERSE OSMOSIS SYSTEMS

The principal contaminants that affect RO systems are iron, manganese, calcium carbonate, calcium sulfate, colloidal matter and microorganisms. Silica in colloidal form is also reported to lower the efficiency of RO Systems. These contaminants foul the membranes in their own characteristic way. Fouling is a general term used to indicate several related but different effects, that may affect membranes. It is used to indicate the trapping or growth of some type of material on the surface of the membrane or within the device itself. All types of membranes and devices are susceptible to fouling but depending on the impurities in water, the type of fouling and degree of severity varies. Accordingly treatment methods also vary. Different types of fouling and treatment procedures are discussed briefly in the following pages.

IRON AND MANGANESE

Iron and manganese are notorious because they cause serious fouling of the membranes. These ions are generally present in the feed water in soluble form. They get oxidized in the system either ahead of the membrane or on the membrane itself. On oxidation, they are converted into insoluble species and deposit over the membranes. Since iron is the common constituent of water, iron fouling is the most prevalent.

Dissolved or suspended iron in the feed water is one of the most damaging foulants. A point need to be mentioned here is that iron fouling occurs rapidly. The fouling is due to oxidation of ferrous iron to ferric iron and subsequent deposition over membranes as ferric hydroxide.

Ground waters generally contain only traces of iron and may not pose a problem. The iron in such waters would be in the ferrous form and soluble in water. As soon as it comes into contact with air, it will be immediately oxidized to ferric iron and gets precipitated. Iron in ground waters may be removed by aeration followed by sand filtration.

Surface water contains iron in significant concentrations. The iron in surface water will normally be present in ferric form (in completely oxidized form and generally in insoluble condition) which may be removed by clarification and / or filtration.

The methods followed at present for the removal of iron and manganese are listed below.

i) *Filtration*
ii) *Aeration, settling and filtration*
iii) *Coagulation, settling and filtration*
iv) *Sodium-cation exchange softening*
v) *Manganese-zeolite treatment*

A knowledge of the form of iron in which it is present would be more useful in deciding the suitable type of treatment.

CALCIUM CARBONATE AND CALCIUM SULFATE

Scaling is the most severe type of fouling that causes the most serious damage to membranes. The salts of the feed water normally tend to deposit when their solubility products are exceeded. While this is the case, the salts in the feed water are usually concentrated 2–4 times, and hence the chances of exceeding their solubility limits are more, which results in their precipitation. Then scale will form on the surface of the membrane (i.e) raw water side of the membrane. Calcium carbonate and calcium sulfate are the chief scale formers. Calcium carbonate scaling can be averted by the addition of hydrochloric or sulfuric acid to the feed water. The quantity of acid to be added depends on the feed water quality. Calcium sulfate scaling is avoided by the addition of sodium hexa - meta phosphate that inhibits the precipitation of calcium sulfate by forming soluble complex.

COLLOIDAL MATTER

Colloidal fouling of the RO membranes occurs due to the entrapment of colloidal particles on the surface of the membrane. The colloidal particles normally found in feed water are natural clays (belonging to aluminium silicate class). Aluminium hydroxide (in colloidal form) is also seen in certain feed waters, the occurrence of which is attributed to improper clarification while using alum as coagulant. Iron colloids are also not uncommon as they are derived as corrosion products from piping and pumps.

Colloidal fouling occurs mainly due to the coagulation of colloidal particles during the reverse osmosis process. Concentration of colloids and stability of colloids are the two factors that determine the magnitude and rate of coagulation. Concentration of colloids in water is usually expressed as "Silt Density Index" (SDI) also called as "Fouling Index". This is determined by passing water through a 0.45μm membrane filter at a pressure of 30 pounds lb/in^2. This is the best available method for the determination of colloids. Usually spiral-wound membranes (cellulose acetate) can tolerate a higher SDI than the hollow fibre membranes (polyamide and cellulose triacetate).

As a thumb rule, water with a SDI less than 4 need not be treated. The SDI of well waters is normally less than 3.0. It does not cause any fouling and needs no treatment. The SDI of surface waters are normally high and range from 10 – 175 and cause serious fouling and hence should be treated.

The recommended method of treatment is filtration. It is applicable to waters with SDI below 10. It is reported that filtration using sand, carbon or other filter media reduce the colloidal concentration substantially. Diatomaceous earth is reported to be the most effective medium. However pilot plant studies are necessary to select the best medium and to find out the degree of reduction of colloidal matter.

When the SDI is very high (more than 50) it is essential to treat the water with coagulants. Alum (aluminium sulfate) is the preferred coagulant. However iron salts (ferric sulfate and chloride) are also proved to be more effective for certain waters. Needless to say that coagulation should be followed by clarification and filtration. In order to secure fast settling, polyelectrolytes are used at present. Non-ionic or anionic polyelectrolytes are more effective.

If RO membranes are affected by colloidal fouling, what is the remedy? The membrane can be cleaned. There are several cleaning solutions available in the market – both common and patented. The common cleaning solutions are citric acid, dilute hydrochloric acid (pH 2.5) and dilute sodium hydroxide (pH 11.0).

MICRO-ORGANISMS

Biological fouling is caused by the growth of micro-organisms. If they grow on membrane, they are likely to block the passage of water. The hollow fibre membranes in particular becomes a good breeding ground for bacteria as this membrane is attacked seriously even by minute quantities of chlorine. Due to this property, chlorinated water is not allowed to permeate through the membrane. In addition to blocking, biological fouling results in the degradation of membranes and life span of membranes would be reduced considerably.

The RO membranes are capable of rejecting nearly 100% of bacteria (and viruses). However, the chances of product water getting contaminated with microbes cannot be precluded. Minute quantities of feed water are likely to enter into the product water through mechanical seals and 'O' – rings. Though the quantum of such feed water entering the product water line is meagre, it is of much concern to pure drinking water production, pharmaceutical industries and electronic industries.

In order to prevent biological fouling and to obtain a sterile product water, the feed water is normally treated with chlorine. However the chlorinated feed water should be dechlorinated thoroughly by passing the chlorinated feed water through activated carbon filter. Otherwise the residual chlorine will attack the membrane materials seriously. It should be borne in mind that polyamide membranes are more vulnerable to chlorine attack and get damaged on prolonged exposure. It cannot withstand even 0.1 mg/l chlorine. Though the cellulose acetate membrane is capable of withstanding to a chlorine concentration upto 1.0 mg/l, it will also be affected seriously. If necessity arises, the membrane may be treated with formaldehyde solution. RO units running continuously need not require such treatment. However in certain instances, for example, after an extended period of shut down, treatment with formaldehyde solution is imperative.

APPENDIX

TABLE – I
SOME OF THE COMMON ABBREVIATIONS

A	angstrom unit	in.	inch
atm	atmosphere	kw	kilowatt
ave	average	l	litre
bbl	barrel	lb	pound
Bé	Baumé degree	m	meter
BOD	Biochemical Oxygen Demand (in 5 days at 20°C, unless otherwise stated)	meq	milliequivalent
Btu	British thermal unit	mg	milligram
C	Centigrade degree	mgad	million gallons per acre per day
cal	calorie	mgd	millions gallons per day
cc	cubic centimeter	min	minute
cfm	cubic foot per minute	ml	milliliter
cfs	cubic foot per second	mm	millimeter
cgs	centimeter-gram-second system	MPN	Most Probable Number
cm	centimeter	$m\mu$	millimicron
csm	cubic foot per second per square mile	μ	micron
cu	cubic	μc	microcurie
deg	degree	μg	microgram
F	Fahrenheit degree	N	normal
fpm	foot per minute	No.	number
fps	foot per second	ppb	part per billion
ft	foot	ppm	part per million
gal	gallon	psi	pound per square inch
gpad	gallon per acre per day	psia	pound per square inch, absolute
gpcd	gallon per capita per day	psig	pound per square inch, gage
gpd	gallon per day	rpm	revolution per minute
gph	gallon per hour	sec	second
gpm	gallon per minute	sq	square
hp	horsepower	wt	weight
hr	hour		

TABLE – II
LENGTH

1 Kilometer	= 1000 meters	1 micron	= 0.000001 meter (10^{-6}m)
1 meter	= 100 centimeters	1 millimicron	= 0.000000001 meter (10^{-9}m)
1 meter	= 1000 millimeters	1 Angstrom	= 0.0000000001 meter (10^{-10}m)
1 millimeter	= 0.03937 inch	1 inch	= 2.540 centimeters
1 centimeter	= 0.3937 inch	1 foot	= 30.480 centimeters
1 meter	= 39.37 inches	1 foot	= 0.3048 meter
1 meter	= 3.2808 feet	1 yard	= 91.440 centimeters
1 meter	= 1.09361 yards	1 yard	= 0.9144 meter
1 kilometer	= 0.6214 mile	1 mile	= 1.6093 kilometers.

TABLE – III
AREA

1 square centimeter	= 0.155 square inch	1 sq. inch	= 6.4516 sq.cms
1 square meter	= 1555.0 sq.inch	1 sq.foot	= 929.0341 sq.cms
1 sq.meter	= 10.764 sq.feet	1 sq.foot	=0.0929 sq.meter
1 sq.meter	= 1.196 sq. yards	1 sq.yard	= 0.8361 sq.meter
1 sq.km	= 0.3861 sq.mile	1 acre	= 43560 sq.ft
1 sq.mile	= 2.590 sq.km	1 acre	= 0.4047 hectare
		1 hectare	= 2.471 acres

TABLE – IV
VOLUME

1 kilo litre	= 1000 litres	1 cubic meter	= 1000 litres
1 deci litre	= 0.1 litre	1 cubic meter	= 35.145 cubic feet
1 centi litre	= 0.01 litre	1 cubic feet	= 1.3079 cubic yard
1 milli litre	= 0.001 litre	1 cubic centimeter	= 0.06102 cubic inch
1 micro litre	= 0.000001 litre	1 cubic inch	= 16.3872 cubic centimeter
1 litre	= 1000 milli litres	1 cubic foot	= 0.02832 cubic meter
		1 cubic yard	= 0.7646 cubic meter

TABLE – V
WEIGHT

1 ton	= 1000 Kilograms	1 microgram	= 0.001 mg (10^{-3} mg or 10^{-6}g)
1 Kilogram	= 1000 grams	1 microgram	= 1000 nanograms (10^{-6} mg or 10^{-9}g)
1 gram	= 1000 milligrams	1 Kilogram	= 2.205 pounds
1 milligram	= 1000 micrograms	1 pound	= 453.6 grams

TABLE – VI
APPLICATION OF METRIC UNITS

Description	Unit	Symbol
Precipitation run-off	milli metre (1 mm of rain = 1 litre/m^2)	mm
River flow	Cubic metre per second	m^3/S
Flow in pipes	Cubic metre per second or litre per second	m^3/S or l/s
Discharges	Cubic metre per day	m^3/d
Usage of water	litre per person per day	l/person/day
Density	Kilogram per cubic metre	Kg/m^3
Concentration	milligram per litre or gram per cubic metre	g/m^3
B.O.D. loading	Kilogram per cubic metre per day	$Kg/m^3/d$
Hydraulic load per unit area	Cubic metre per square-metre per day	$m^3/m^2/d$
Hydraulic load per unit volume	Cubic metre per cubic metre per day	$m^3/m^2/d$
Velocity	metre per second	m/s

TABLE – VII
VOLUME CONVERSION FACTORS

From to	*cm^3*	*liters*	*m^3*	*in^3*	*ft^3*	*yd^3*	*gal (U.S)*	*gal (U.K)*
cm^3	1	0.001	1×10^{-6}	0.06102	353×10^{-5}	1.3×10^{-6}	2.64×10^{-4}	2.2×10^{-4}
litre	1000	1	0.001	61.02	0.03532	0.00131	0.2642	0.220
m^3	1×10^{-6}	1000	1	6.1×10^{4}	35.31	1.308	264.2	220
in^3	16.39	0.01639	1.64×10^{-5}	1	5.79×10^{-4}	2.14×10^{-5}	0.00433	0.00360
ft^3	283×10^{4}	28.32	0.02832	1728	1	0.03704	7.481	6.229
yd^3	7.65×10^{5}	764.5	0.7646	4.67×10^{4}	27	1	202.0	168.2
gal (U.S)	3785	3.785	0.00379	231.0	0.1337	0.00495	1	0.8327
gas (U.K)	4546	4.546	0.00455	277.4	0.1605	0.00595	1.201	1

TABLE - VIII
FLOW RATE CONVERSION FACTORS

From to	*litre/s*	*gal/min*	*ft^3/s*	*ft^3/min*	*m^3/s*	*m^3/min*
litre/s	1	15.85	0.03532	2.119	0.001	0.06
gal/min	0.06309	1	0.00223	0.1337	0.000063	0.00379
ft^3/s	28.32	448.8	1	60	0.02832	1.699
ft^3/min	0.4719	7.481	0.01667	1	0.000472	0.02832
m^3/s	1000	15852	35.31	2118.6	1	60
m^3/min	16.667	264.2	0.5885	35.31	0.01667	1

TABLE – IX
CONVERSION FACTORS FOR TEMPERATURES SCALES

The following formula may be used to convert temperatures from one scale to another

	Temperature wanted in			
Temp. given in	**^{0}C**	**^{0}K**	**^{0}F**	**^{0}Rank**
^{0}C	C	C + 273.16	1.86 C + 32	1.8C + 491.4
^{0}K	K – 273.16	K	1.8K – 459.4	1.8K
^{0}F	0.556F – 17.8	0.556F + 255.3	F	F + 459.4
^{0}Rank	0.556 Rank – 273.1	0.556 Rank	Rank – 459.4	Rank

Table - X
CONVERSION FACTORS FOR GASES

Multiply	By	To obtain
p. p. m. of oxygen	0.698	ml/l
ml/l of oxygen	1.43	p.p.m
p.p.m of carbon dioxide	0.506	ml/l
ml/l of carbon dioxide	1.98	p.p.m

TABLE - XI
SIEVE MESH CONVERSION TABLE
SCREEN EQUIVALENTS

Sieve No.	U.S. Standard Opening		Meshes per Inch	Tyler Standard Opening		Meshes per Inch	British Standard Opening	
	mm	Inches		mm	Inches		Mm	Inches
12	1.68	0.0661	10	1.65	0.065	10	1.68	0.0660
14	1.41	0.0555	12	1.40	0.055	12	1.40	0.0553
16	1.19	0.0469	14	1.17	0.046	14	1.20	0.0474
18	1.00	0.0394	16	0.991	0.039	16	1.00	0.0395
20	0.84	0.0331	20	0.833	0.0328	18	0.853	0.0336
25	0.71	0.0280	24	0.701	0.0276	22	0.669	0.0275
30	0.59	0.0232	28	0.589	0.0232	25	0.599	0.0236
35	0.50	0.0197	32	0.495	0.0195	30	0.500	0.0197
40	0.42	0.0165	35	0.417	0.0164	36	0.422	0.0166
45	0.35	0.0138	42	0.351	0.0138	44	0.353	0.0139
50	0.297	0.0117	48	0.295	0.0116	52	0.295	0.0116
60	0.250	0.0098	60	0.246	0.0097	60	0.251	0.0099
70	0.210	0.0083	65	0.208	0.0082	72	0.211	0.0083
80	0.177	0.0070	80	0.175	0.0069	85	0.178	0.007
100	0.149	0.0059	100	0.147	0.0058	100	0.152	0.006
200	0.074	0.0029	200	0.074	0.0029	200	0.076	0.003
325	0.044	0.0017	325	0.043	0.0017	240	0.066	0.0026

TABLE - XII
INTERNATIONAL ATOMIC WEIGHTS 1973

Element	Symbol	Atomic No.	Atomic Weight
Actinium	Ac	89	(227)
Aluminium	Al	13	26.9815
Americium	Am	95	(243)
Antimony	Sb	51	121.75
Argon	Ar	18	39.948
Arsenic	As	33	74.9216
Astaline	At	85	(210)
Barium	Ba	56	137.34
Berkeylium	Bk	97	(247)
Beryllium	Be	4	9.0122
Bismuth	Bi	83	208.9804
Boron	B	5	10.81
Bromine	Br	35	79.904
Cadmium	Cd	48	112.40
Calcium	Ca	20	40.08
Californium	Cf	98	(251)
Carbon	C	6	12.011
Cerium	Ce	58	140.12
Cesium	Cs	55	132.9054
Chlorine	Cl	17	35.453
Chromium	Cr	24	51.996
Cobalt	Co	27	58.9332
Copper	Cu	29	63.542
Curium	Cm	96	(247)
Dysprosium	Dy	66	162.50
Einsteinium	Es	99	(254)
Erbium	Er	68	167.26
Europium	Eu	63	151.96
Fermium	Fm	100	(257)
Fluorine	F	9	18.9984
Francium	Fr	87	(223)
Godolinium	Gd	64	157.25
Gallium	Ga	31	69.72
Germanium	Ge	32	72.59
Gold	Au	79	196.9665
Hafnium	Hf	72	178.49
Helium	He	2	4.0026
Holmium	Ho	67	164.9304

Element	Symbol	Atomic No.	Atomic Weight
Hydrogen	H	1	1.0079
Indium	In	49	114.82
Iodine	I	53	126.9045
Iridium	Ir	77	192.22
Iron	Fe	26	55.847
Krypton	Kr	36	83.80
Lanthanum	La	57	138.9066
Lawrencium	Lr	103	(260)
Lead	Pb	82	207.0
Lithium	Li	3	6.941
Lutetium	Lu	71	174.97
Magnesium	Mg	12	24.305
Manganese	Mn	25	54.9380
Mendelevium	Md	101	(258)
Mercury	Hg	80	200.59
Molybdenum	Mo	42	95.94
Neodymium	Nd	60	144.24
Neon	Ne	10	20.179
Neptunium	Np	93	237.0482
Nickel	Ni	28	58.70
Niobium	Nb	41	92.9064
Nitrogen	N	7	14.0067
Nobelium	No	102	(255)
Osmium	Os	76	190.2
Oxygen	O	8	15.9994
Palladium	Pd	46	106.4
Phosphorus	P	15	30.9738
Platinum	Pt	78	195.09
Plutonium	Pu	94	(244)
Polonium	Po	84	(209)
Potassium	K	19	39.098
Praseodymium	Pr	59	140.9077
Promethium	Pm	61	(145)
Protactinium	Pa	91	231.0359
Radium	Ra	88	226.0254
Radon	Rn	86	(222)
Rhenium	Re	75	186.207
Rhodium	Rh	45	102.9055
Rubidium	Rb	37	85.4678
Ruthenium	Ru	44	101.07
Samarium	Sm	62	150.4

Element	Symbol	Atomic No.	Atomic Weight
Scandium	Sc	21	44.9559
Selenium	Se	34	78.96
Silicon	Si	14	28.086
Silver	Ag	47	107.868
Sodium	Na	11	22.9898
Strontium	Sr	38	87.62
Sulphur	S	16	32.06
Tantalum	Ta	73	180.9479
Technetium	Tc	43	(97)
Tellurium	Te	52	127.60
Teribium	Tb	65	158.9254
Thallium	Tl	81	204.37
Thorium	Th	90	232.0381
Thulium	Tm	69	168.9342
Tin	Sn	50	118.69
titanium	Ti	22	47.90
tungsten	W	74	183.85
Uranium	U	92	238.029
Vanadium	V	23	50.9414
Xenon	Xe	54	131.30
Ytterbium	Yb	70	173.04
Yttrium	Y	39	88.9059
Zinc	Zn	30	65.38
Zirconium	Zr	40	91.22

TABLE – XIII
DATA ON CHEMICALS USED IN WATER AND WASTE WATER TREATMENT

Chemical Name, Formula and Formula Weight	Common Name/Trade Name	Available form	Commercial strength	Characteristic	Solubility gm/100 ml water				Weight/ lb/cu.ft	Storage space tone cu.ft/ton	Shipping containers	Suitable handling materials
					0°C	10°C	20°C	30°C				
Activated Carbon – See Carbon activated												
Aluminium sulfate $Al_2(SO_4)_3.14H_2O$ 594	Alum Filter Alum Sulfate of Alumina	Ivory/Brown colored slab, lump, granule, powder, Also available in liquid form	14.5% to 17.5% Al_2O_3	pH of 1% solution 3.4	60.8	65.3	71.0	78.8	Slab lump (or) ground 57-67 powder-ed 38-45	30-35 45-53	Bags, barrels and in bulk; liquid is transported by Tankers	Dry : Iron, steel Liquid: Lead lined rubber, silicon asphalt

Chemical Name, Formula and Formula Weight	Common Name/Trade Name	Available form	Commercial strength	Characteristic	Solubility gm/100 ml water				Weight/ lb/cu.ft	Storage space tone cu.ft/ton	Shipping containers	Suitable handling materials
					0°C	10°C	20°C	30°C				
Alum, ammonium $Al_2(SO_4)_3.(NH_4)_2.SO_4.2H_2O$ 906	Ammonia Alum (Sometimes called as crystal alum)	Lump, nut, powder	11% Al_2O_3	pH of 1% solution 3.5	3.9	9.5	15.1	20.0	64-68	30-31	Bags, barrels, boxes, Cartons, Kegs and bulk	Lead lined rubber, silicon Iron, stone ware.
Alum potassium $Al_2(SO_4)_3.K_2SO_4.24H_2O$ 949	potash alum	lump, crystal	10.7% Al_2O_3	pH 1% solution 3.5	5.7	7.6	11.4	16.6	64-68	30-31	Bags, barrels, boxes, cartons, and kegs	Lead lined rubber, stone ware.
Bentonite	colloidal clay, volclay, wilkinite	Powder, pallet, mixed sizes	-	-	-	-	-	-	60	-	Bags bulk	Iron steel
Bleaching power See "chlorinated lime"												
Calcium Carbonate $CaCO_3$ 100	Lime stone, Calcite, Chalk, whiting	Crushed and graded granules; also available in powder form	-	-	-	-	13	-	Granule 100-115 powder 48-71	18-20 28-42	Bags, barrels bulk	
Calcium chloride $CaCl_2$ 111	-	Flake: 77-78% $CaCl_2$ Solid: 73-75% $CaCl_2$ Liquid: 40% $CaCl_2$	-	-	59.5	65.0	74.5	102	60-66	29-31	Flakes: moisture proof bags or drums Solids: drums	-
Calcium hydroixed $Ca(OH)_2$ 74.1	Hydrated lime, slaked lime	Powder (light and dense)	85-99% $Ca(OH)_2$	Hopper agitation required for dry feed of light form.	0.18	0.17	0.16	0.15	25-50	40-80	Bags, barrels, bulk	Rubber, Iron, Steel, cement and asphalt
Calcium hypochlorite $CaOCl_2.4H_2O$ 199	"ATH" "Perchloron" "Pittchlor"	Powder	70% available chlorine	- Generaly 1-3% available Cl_2 solution used.	Highly soluble				-	-	Cans and drums	Glass, rubber, stoneware and wood

Chemical Name, Formula and Formula Weight	Common Name/Trade Name	Available form	Commercial strength	Charact-eristic	Solubility gm/100 ml water				Weight/ lb/cu.ft	Storage space tone cu.ft/ton	Shipping containers	Suitable handling materials
					0°C	10°C	20°C	30°C				
Calcium oxide CaO 56.1	lime, burnt lime, chemical lime, quick lime, unslaked lime	Lump, pebble ground and powder.	75 to 99% CaO	Slakes with water to form hydrated lime $Ca(OH)_2$ For slaking 1 Kg. lime, 10 litres of water is required	–	–	–	–	Lump 50-65 Ground 50-70 Pulv. 39-71	31-40 29-40 28-52	Bags, barrels and in bulk	Rubber, Iron, steel, cement and asphalt
Calcium sulfate $CaSO_4.2H_2O$ 172	Gypsum	Ground	–	–	0.18	0.19	0.20	0.21	50-60	33-40	Bags, barrels and in bulk	–
Carbon-activated C 12	–	Black granules and powder	–	–	Insoluble				Granular 12-30 Powder 15-28	67-167 72-134	Bags, bulk	Dry iron, steel, wet-rubber, stainless steel.
Caustic soda See Sodium-hydroxide												
Chlorinated lime $CaO.2CaOCl_2.3H_2O$ 364	Bleaching Powder, chloride of lime	White powder	25-37% available chlorine	Deteriorates on exposure to air	Very soluble				48	–	Drums	Glass, rubber, stone ware, wood
Chlorine Cl_2 70.9	–	Chlorine gas Liquid chlorine (liquid field gas under pressure in steel cylinders	99.8% Cl_2	–	1.46	0.980	0.716	0.562	–	–	Cylinders Tank ers, Tank trucks	Dry Iron Copper, steel, wet gas – glass, hard rubber.
Chlorine dioxide ClO_2 102.9	Chlorine dioxide	Yellow-red gas	26.3% available chlorine	–	–	–	–	–	–	–	Generated at the site itself	Plastic, soft rubber (Hard rubber to be avoided)
Copper sulfate $CuSO_4.5H_2O$ 249.6	Blue vitriol Blue stone	Crystal, lump powder	99% $CuSO_4$	–	19.2	–	26.4	31.2	Crystal 75-90 lump 73-80 Powder 60-64	–	Bags, barrels, drums	Iron, Stainless steel, Silicon, asphalt
Disodium phosphate (see Sodium phosphate) Dolomiticlime $CaO + MgO$ (Content of MgO varies)	Dolomitic lime, High magnesium lime	Lump, pebble, ground or powder	58% CaO and 40% MgO	Typical for Silica removal Slakes with water forming $Ca(OH)_2$+ MgO the later Slakes very slowly	–	–	–	–	Lump 50-65 Pebble 60-65 Ground 50-75 Pulv 37-63	31-40 31-34 27-40 32-54	Bags, barrels and in bulk	Wood, Iron steel

Chemical Name, Formula and Formula Weight	Common Name/Trade Name	Available form	Commercial strength	Charact-eristic	Solubility gm/100 ml water				Weight/ lb/cu.ft	Storage space tone cu.ft/ton	Shipping containers	Suitable handling materials
					0°C	10°C	20°C	30°C				
Dolomitic lime, hydrated $Ca(OH)_2$ + MgO content of MgO varies	Dolomitic lime, hydrated	Powder	62% $Ca(OH)_2$ and 32% MgO	Typical for silica removal	–	–	–	–	28-52	39-72	Bags, barrels and in bulk	Wood, Iron Steel
Ferric chloride (i) $FeCl_3$ (ii) $FeCl_3.6H2O$ (iii) $FeCl_3$ (35-45%) solution	Anhydrous ferric chloride Crystal ferric Chloride "Ferrichlor" Chloride of Iron	Green black powder Yellow brown lump Dark brown Syrupy liquid	98% $FeCl_3$ 34% Fe 59-61% $FeCl_3$ 20-21% Fe 37-47% $FeCl_3$ 20-21% Fe	Highly hygroscopic; lumps and Powder should be stored in air tight containers.	Completely soluble				–	–	Casks, kegs barrels car-boys trucks, tankers.	Rubber, glass, stoneware synthetic-resins
Ferric Sulfate $Fe_2(SO_4)_3$ 400	Ferrisul Ferriflac	Granules Reddish brown powder	Composition from different sources varies 70%-90% $Fe_2(SO_4)_3$ some grades contain $Al_2(SO_4)_3$ also	Mildly hygroscopic coagulated at pH 8.5 – 11.0	Highly soluble, to dissolve in cold water, add 2 parts water to 1 part ferric sulfate with stirring.				60-70	29-34	Bags, barrels, kegs, drum and in bulk	Rubber, plastic, lead ceramic, stainless-steel
Ferrous sulfate $FeSO_4.7H_2O$ 278	Copper as green vitriol	Crystals, granules	55% $FeSO_4$ 20% Fe	Hygoscopic: cakes on storage. Optimum pH 8.5 – 11.0	28.7	37.5	48.5	60.2	63-66	30-32	Bags, barrels kegs and in bulk	Wood, lead asphalt, concrete
Hydrated lime (See Calcium hydroxide)												
Lime (See Calcium oxide)												
Hydrazine N_2H_4 32.05	–	Colorless oily liquid	–	Powerful reducing agent fuming in air penetrating odor resembling that of ammonia	Miscible with water				–	–	Glass	Glass
Hydrochloric acid HCl 36.45	Muriatic acid	sometime-es colored by traces of iron chlorine and organic matter	38% HCl	Fumes in air pH of 1N HCl is 0.1 0.1N HCl = 1.1 0.01N HCl = 2.02	–	–	–	–	–	–	Bulk in Tankers	Plastic rubber

Chemical Name, Formula and Formula Weight	Common Name/Trade Name	Available form	Commercial strength	Charact-eristic	Solubility gm/100 ml water				Weight/ lb/cu.ft	Storage space tone cu.ft/ton	Shipping containers	Suitable handling materials
					0°C	10°C	20°C	30° C				
Meta Sodium-phosphate (See sodium phosphate meta)	–	–	–									
Soda ash (See Sodium carbonate)												
Sodium aluminate $Na_2O.Al_2O_3$ 164	Soda alum	Brown powder liquid (27°Be)	70-80% $Na_2Al_2O_4$ Min. 32% $Na_2.Al_2O_4$	For dry feed hopper agitation required	Complete				50-60	–	Bags, drums tankers	Iron, plastic, rubber, steel
Sodium carbonate Na_2CO_3 106	Soda ash	White powder Extra light, dense	Light soda ash is used in water treatment. 99.4% Na_2CO_3 58% Na_2O	pH of 1% solution 11.3 Hopper agitation required for dry feeding	7.0	12.5	4.5	38.8	23 35 65	–	Bags, barrels bulk [Carloads, trucks]	Iron, rubber, steel
Sodium chloride NaCl 58.5	Salt, common salt, rock salt evaporated salt	Crystal powder	98% NaCl	–	35.7	35.8	36.0	36.3	50-70	29-40	Bags, barrels, bulk	Rubber, cement, plastic, bronze
Sodium hydroxide NaOH 40	Caustic soda	Flakes, solid ground, liquid	98.06 NaOH (76% Na_2O) 50-70% solution	Solid Highly hygoscopic pH of 1% solution 12.9	42.0	51.5	109	119	–	–	Flake : drums Solid : drums Solution: in tankers	Cast iron steel, rubber.
Sodium hypochlorite NaOCl 74.5	Sodium hypochlorite	Light yellow liquid	12-15% available chlorine	–	–	–	–	–	–	–	Carboys, Bulk in tankers	Glass plastics, rubber, ceramics
Sodium hexameta phosphate $(NaPO_3)_6$ 750	'Calgon' glassy phosphate vitreous phosphate	crystal, flakes powder	66% P_2O_5 (unadjusted)	pH of 0.25% solution 6.0-8.3	–	–	–	–	47	–	Bags	Plastics, hard rubber, stainless steel
Sodium phosphate, mono NaH_2PO_4 (anhydrous) 120 $NaH_2PO_4.H_2O$ 138	–				Monohydrate 66.6	80.4	98.0	12.3	Anhydrous 55-70 Mono hydrate 58-79	29-37 26-35	Bags, barrels, boxes, kegs drums	Cast iron, steel
Sodium phosphate, Disodium phosphate, Disodium hydrogen phosphate Na_2HPO_4 – 142 (anhydrous) $Na_2HPO_4.12H_2O$ - 358	Basic sodium phosphate DSP secondary sodium phosphate	Crystal	19.5% P_2O_5	Precipitates calcium and magnesium pH of 1% solution 9%	Anhydrous 1.7 Do decahydrate 4.2	3.6 8.9	7.7 19.3	20.8 52.4	Anhydrous 53-62 46-53 Do deca hydrate	32-39 38-44	Bags, barrels, kegs drum	Cast Iron, steel

Chemical Name, Formula and Formula Weight	Common Name/Trade Name	Available form	Commercial strength	Charact-eristic	Solubility gm/100 ml water				Weight/ lb/cu.ft	Storage space tone cu.ft/ton	Shipping containers	Suitable handling materials
					0°C	10°C	20°C	30°C				
Sodium phosphate Trisodium phosphate $Na_3PO_4.H_2O$ 182 $Na_3PO_4.12H_2O$ 380	Normal sodium phosphate Tertiary sodium phosphate	Crystal coarse Medium standard	19% P_2O_5	pH of 1% solution 11.9	Monohydrate						Bags, barrels boxes kegs, drum	Cast Iron, steel
					1.7	4.6	12.3	22.2	83-90	22-24		
					Do deca hydrate							
					3.5	9.5	25.5	46.4	56-60	32-36		
Sodium tripoly phosphate (STPP) polypenta sodium triphosphate $Na_5P_3O_{10}$ – 367.9	–	slightly hygroscopic granules	–	pH of 1% solution 9.7 to 9.8	20 gm at 25°C and 86.5 gm at 100°C				–	–	–	–
Sodium Silicate $Na_2O.SiO_2$ (variable ratio of Na_2O to SiO_2)	Water glass Silicate of soda	opaque viscous liquid	38° – 40° Be solution	pH of 1% solution 12.3	Miscible in water in all proportions				86.1	–	Drums Tankers	Cast iron rubber, steel
Sodium sulfite Na_2SO_3 126	Sulfite	White crystals powder	23% SO_2	Sulfurous taste and odor	14.0	20.0	27.0	36.0	80-91	23-26	Bags, barrels Kegs and drums	Iron, steel wood
Sulfur dioxide SO_2 64.1	Sulfurous acid anhydride	Colorless gas (liquefied gas under pressure in steel cylinders)	99% SO_2	Highly corrosive and irritating gas	22.8	16.2	11.3	7.8	–	–	Steel cylinders, Tankers	Aluminium, brass, stainless steel
Sulfuric acid H2SO4 98.1	Oil of vitriol	Liquid	98% 93.2% (66° Be) 77.7% (60° Be)	Approximate pH value of 0.5% solution 1.2	Miscible with water in all proportions				–	–	Bottles, carboys, drums, tankers	Concentr-ated: Iron, steel, Dilute: glass, lead, porcelain rubber.
Sodium – tripolyphosfate (See above – after sodium phosphate – meta)												
Tetra Sodium phosphate Pyro $Na_4P_2O_7.10H_2O$ 445.8	Alkaline sodium pyrophosphate	White powder	53% P_2O_5	pH of 1% solution 10.5	–	–	–	0.6	68	–	Bags barrels, Kegs	Cast Iron, steel.

INDEX

A

B

C

REFERENCES

1. H. De. La. Coux, "INDUSTRIAL USES OF WATER", 1903 (Translated from the French and Revised by Arthen Morris) Scott, Greenwood & Son, London
2. Harald .A. Leverin, "INDUSTRIAL WATERS OF CANADA", 1942, Edmond Cloutier, Ottawa, Canada
3. Edwin Windle Taylor, "THE EXAMINATION OF WATERS AND WATER SUPPLIES", 6th Edition, 1949, J & A Churchill Ltd.
4. Eskell Nordell, "WATER TREATMENT FOR INDUSTRIAL AND OTHER USES", 1951, reinhold Publishing Corporation, New York, USA.
5. Fair & Geyer, "WATER SUPPLY AND WASTE WATER DISPOSAL", 1954, John Wiley & Sons, New York.
6. Louis Klein, "River Pollution II Causes and Effects", 1962, Butterworths, London
7. James A. Kent (Ed), "REIGEL'S INDUSTRAIL CHEMISTRY", 1962, Reirhold Publishing Corporation, U.S.A
8. Leo. I. Pincus "PRACTICAL BOILER WATER TREATMENT" 1962, McGraw-Hill Book Co., New York.
9. J.E. McKee & H.W. Wolf, "WATER QUALITY CRITERIA" 2nd Edition, 1963, The Resources Agency of California State Water Resources Control Board, USA.
10. K.S.Spiegler (Ed), "Principles of Desalination", 1966, Academic Press, Newyork.
11. American Society for Testing and Materials, "A MANUAL ON WATER" ASTM Special Technical Publication No. 442, 1969, ASTM-Philadelphia.
12. American Water Works Association, "WATER QUALITY AND TREATMENT", 3rd Edition, 1971, McGraw-Hill Book Company, New York.
13. D.M.Considine, "Chemical and Process Technology Encyclopedia", 1974, McGraw Hill Book Company, New York.
14. R. N. Shreeve and Joseph .A Brink .Jr, "CHEMICAL PROCESS INDUSTRIES", 4th Edition, 1977, McGraw-Hill International Book Company, New York.

15. C.N.Sawyer and P.L. McCarty, "Chemistry for Environmental Engineering", 1978, McGraw Hill Book Company, New York.

16. Pelczar, Reid and Chan, "MICROBIOLOGY", 4th Edition, 1978, Tata McGraw-Hill Publishing Company Limited, New Delhi.

17. N. F. Voznaya, "CHEMISTRY OF WATER AND MICROBIOLOGY", 1981, Mir Publishers, Moscow.

18. James W.McCoy, "The Chemical Treatment of Boiler Water", 1981, Chemical Publishing Co., New York.

19. Walter Lorch (Ed), "HAND BOOK OF WATER PURIFICATION", 1981, Mc-Graw Hill Book Company (UK) Ltd., London.

20. Frank N. Kemmer (Ed), "THE NALCO WATER HAND BOOK", 2nd Edition, 1988, Mc-Graw Hill Book Co., New York.

21. Osmonics, "PURE WATER HAND BOOK", 1991, Osmonics Inc., Minnetonka, USA.

22. Americal Society of Mechanical Engineers, "Consensus on Operating Practice for the Control of Feedwater and Boiler Water Chemistry in Modern Industrial Boilers", 1994, ASME.

23. George S. Solt and Chris B. Shirley, "AN ENGINEERS' GUIDE TO WATER TREATMENT", 1991, Avebury Technical, England.

24. N. Manivasakam, "WATER USED IN TEXTILE PROCESSING, QUALITY TREATMENT AND ANALYSIS", 1995, Sakthi Publications, Coimbatore (TN). India.

25. World Health Organization, "Guidelines for Drinking Water Quality, Volume–2, Health Criteria and Supporting Information", 1996 WHO, Geneva, Switzerland.

26. World Health Organization, Guidelines for Drinking Water Quality, Voluem - 3, Surveillance and Control of Community Supplies", 2nd Edition, 1997, WHO, Geneva, Switzerland.

27. American Public Health Association, American Water Works Association, Water Environment Federation, "STANDARD METHODS FOR THE EXAMINATION OF WATER AND WASTE WATER", 20th Edition, 1998, APHA, Washington.

28. Hach Company, "WATER ANALYSIS HAND BOOK", 4th Edition, Revision 2, 2003, Hach Company, Colorado, USA.

29. World Health Organization, "Guidelines for Drinking Water Quality, Volume-1, Recommendations", 3rd Edition, 2004, WHO, Geneva, Switzerland.

www.ingramcontent.com/pod-product-compliance
Lightning Source LLC
LaVergne TN
LVHW062148200926
848484LV00007B/1878